LIVING BY Chemistry
First Edition

Unit 2: Smells Teacher Guide
Molecular Structure and Properties

Angelica M. Stacy
Professor of Chemistry
University of California at Berkeley

with

Janice A. Coonrod
Senior Writer and Developer

Jennifer Claesgens
Curriculum Developer

Key Curriculum Press

Editors	Ladie Malek, Jeffrey Dowling
Project Administrators	Elizabeth Ball, Rachel Merton, Janis Pope
Consulting Editors	Heather Dever, Joan Lewis, Andres Marti
Editorial Advisor	Casey FitzSimons
Production Editor	Andrew Jones
Editorial Production Supervisor	Kristin Ferraioli
Copyeditor	Mary Roybal
Senior Production Coordinator	Ann Rothenbuhler
Production Director	Christine Osborne
Text Designer	Roy Neuhaus Design
Compositor	Precision Graphics
Art Editor	Maya Melenchuk
Illustrators	Ken Cursoe, Greg Hargreaves, Tom Ward
Technical Artist	Precision Graphics
Photo Researcher	Laura Murray Productions
Cover Designer	Diana Ghermann
Prepress and Printer	Sheridan Books, Inc.
Textbook Product Manager	Tim Pope
Executive Editor	Josephine Noah
Publisher	Steven Rasmussen

This material is based upon work supported by the National Science Foundation under award number 9730634. Any opinions, findings, and conclusions or recommendations expressed in this publication are those of the author and do not necessarily reflect the views of the National Science Foundation.

© 2010 by Key Curriculum Press. All rights reserved.

No part of this publication may be reproduced, stored in a retrieval system, or transmitted, in any form or by any means, electronic, photocopying, recording, or otherwise, without the prior written permission of the publisher.

®Key Curriculum Press is a registered trademark of Key Curriculum Press.

The sciLINKS® service includes copyrighted materials and is owned and provided by the National Science Teachers Association. All rights reserved.

Photo Credits—ix: Ken Karp Photography; xii: Ken Karp Photography; xviii: Visuals Unlimited; 83: Ken Karp Photography

Key Curriculum Press
1150 65th Street
Emeryville, CA 94608
editorial@keypress.com
www.keypress.com

Printed in the United States of America
10 9 8 7 6 5 4 3 2 1   15 14 13 12 11 10 09
ISBN: 978-1-55953-989-0

Reviewers and Field Testers

Science Content Advisor

Dr. A. Truman Schwartz, *Macalester College (emeritus), St. Paul, MN*

Teaching and Content Reviewers

Scott Balicki
Boston Latin School
Boston, MA

Greg Banks
Urban Science Academy
West Roxbury, MA

Randy Cook
Tri County High School
Howard City, MI

Thomas Holme
University of Wisconsin-Milwaukee
Milwaukee, WI

Mark Klawiter
Deerfield High School
Deerfield, WI

Carri Polizzotti
Marin Catholic High School
Larkspur, CA

Matthew Vaughn
Burlingame High School
Burlingame, CA

Rebecca Williams
Richland College
Dallas, TX

Field Testers

Carol de Boer
Amador Valley High School
Pleasanton, CA

Wayne Brock
Life Learning Academy
San Francisco, CA

Janie Burkhalter,
Coronado High School
Lubbock, TX

Karen Chang
The Calhoun School
New York, NY

Elizabeth Christopher
El Camino High School
Woodland, CA

Mark Crown
Gateway High School
San Francisco, CA

Susan Edgar-Lee
Hayward High School
Hayward, CA
and
Livermore High School
Livermore, CA

Melissa Getz
Tennyson High School
Hayward, CA

Shannon J. Halkyard
Stuart Hall High School
San Francisco, CA

David Hodul
De La Salle High School
Concord, CA
and
Bishop O'Dowd High School
Oakland, CA

Field Testers (continued)

Kim D. Johnson
*Thurgood Marshall Academic
High School*
San Francisco, CA

Evy Kavaler
Berkeley High School
Berkeley, CA

Bruce Leach
Hill Country Christian School
Austin, TX

Tatiana Lim
Morse High School
San Diego, CA

Kathleen Markiewicz
Boston Latin School
Boston, MA

Steve Maskel
Hillsdale High School
San Mateo, CA

Mardi Mertens
Berkeley High School
Berkeley, CA

Nicole Nunes
*Thurgood Marshall Academic
High School*
San Francisco, CA
and
De La Salle High School
Concord, CA

Tracy A. Ostrom
Skyline High School
Oakland, CA

Pru Phillips
Crawfordsville High School
Crawfordsville, IN

Daniel Quach
Berkeley High School
Berkeley, CA

Carissa Romano
Hayward High School
Hayward, CA

Sally Rupert
Assets High School
Honolulu, HI

Geoff Ruth
Leadership High School
San Francisco, CA

Maureen Wiser
Emery Secondary School
Emeryville, CA

Audrey Yong
*Thurgood Marshall Academic
High School*
San Francisco, CA

Acknowledgments

A number of individuals joined the project as developers for various periods of time along the way to completing this work. Thanks go to these individuals for their contributions to the unit development: Karen Chang, David Hodul, Rebecca Krystyniak, Tatiana Lim, Jennifer Loeser, Evy Kavaler, Sari Paikoff, Sally Rupert, Geoff Ruth, Nicci Nunes, Gabriela Waschewski, and Daniel Quach.

David R. Dudley contributed original ideas and sketches for some of the wonderful cartoons interspersed throughout the book. His sketches provided a rich foundation for the art manuscript.

This work would not have been possible without the thoughtful feedback and great ideas from numerous teachers who field-tested early versions of the curriculum. Thanks go to these teachers and their students: Carol de Boer, Wayne Brock, Susan Edgar-Lee, Melissa Getz, David Hodul, Richard Kassissieh, Tatiana Lim, Evy Kavaler, Geoff Ruth, Nicci Nunes, Gabriela Waschewski, and Daniel Quach.

Dr. A. Truman Schwartz provided a thorough and detailed review of the manuscript. We appreciate his insights and chemistry expertise.

Ladie Malek and Jeffrey Dowling served as the developmental editors for the project, giving feedback and advice.

About the Author

Angelica Stacy is the author and lead developer of *Living By Chemistry*. In addition to her research and publications in materials, physical, and inorganic chemistry, she has distinguished herself as an outstanding educator, receiving numerous awards and honors in education and holding the President's Chair for Teaching at the University of California from 1993 to 1996. Dr. Stacy is a creative and enthusiastic educator whose interest in developing a high school chemistry curriculum arose out of her desire to help students attain a better understanding of the principles and concepts in chemistry. In 2005, the National Science Foundation named Dr. Stacy a Distinguished Teaching Scholar.

Contents
Unit 2 Smells: Molecular Structure and Properties

Introduction to the *Living By Chemistry* Program	viii
Features of the Teacher Guide	xiii
Content Coverage Chart	xvi
Introduction to Smells: Molecular Structure and Properties	xviii
Pacing Guides	xx

SECTION I Speaking of Molecules 1

Lesson 1 Sniffing Around: *Molecular Formulas*	2
Lesson 2 Molecules in Two Dimensions: *Structural Formulas*	10
Lesson 3 HONC if You Like Molecules: *Bonding Tendencies*	18
Lesson 4 Connect the Dots: *Lewis Dot Symbols*	24
Lesson 5 Eight Is Enough: *Octet Rule*	32
Lesson 6 Where's the Fun?: *Functional Groups*	39
Lesson 7 Create a Smell: *Ester Synthesis*	48
Lesson 8 Making Scents: *Analyzing Ester Synthesis*	55

SECTION II Building Molecules 63

Lesson 9 New Smells, New Ideas: *Ball-and-Stick Models*	64
Lesson 10 Two's Company: *Electron Domains*	73
Lesson 11 Let's Build It: *Molecular Shape*	82
Lesson 12 What Shape Is That Smell?: *Space-Filling Models*	91
Lesson 13 Sorting It Out: *Shape and Smell*	100
Lesson 14 How Does the Nose Know?: *Receptor Site Theory*	108

SECTION III Molecules in Action 115

Lesson 15 Attractive Molecules: *Attractions Between Molecules*	116
Lesson 16 Polar Bears and Penguins: *Electronegativity and Polarity*	125
Lesson 17 Thinking (Electro)Negatively: *Electronegativity Scale*	136
Lesson 18 I Can Relate: *Polar Molecules and Smell*	144
Lesson 19 Sniffing It Out: *Phase, Size, Polarity, and Smell*	153

SECTION IV **Molecules in the Body**   161

    Lesson 20 Mirror, Mirror: *Mirror-Image Isomers*   162

    Lesson 21 Protein Origami: *Amino Acids and Proteins*   170

    Lesson 22 Who Nose?: *Unit Review*   180

Introduction to the *Living By Chemistry* Program

Living By Chemistry is a full-year high school chemistry curriculum that meets and exceeds state and national standards. It consists of six teacher guides, a student textbook, kits, and other print and online teaching resources. The teacher guides are central to the curriculum and provide detailed daily lesson plans. The textbook is accessible and highly visual, and follows the sequencing and flow of what happens in the classroom, directly supporting and reviewing the daily learning.

The curriculum consists of six units, each organized around a specific body of chemistry content and a theme that students can relate to. Most units consist of around 20 lessons of 45-minute duration, which can be combined for 90-minute block periods.

Unit 1	Alchemy	Matter, Atomic Structure, and Bonding	28 Lessons
Unit 2	Smells	Molecular Structure and Properties	22 Lessons
Unit 3	Weather	Phase Changes and Behavior of Gases	20 Lessons
Unit 4	Toxins	Stoichiometry, Solution Chemistry, and Acids and Bases	27 Lessons
Unit 5	Fire	Energy, Thermodynamics, and Oxidation-Reduction	20 Lessons
Unit 6	Showtime	Reversible Reactions and Chemical Equilibrium	8 Lessons

A Thematic Approach

A theme-based curriculum captures students' interest, helps them make connections, and improves retention of concepts. It also serves another purpose—it helps to ground the study of chemistry in the natural world and everyday life. Too often, students view chemistry as an inaccessible discipline centered around synthetic chemicals invented in a lab. In reality, chemical processes occur in our bodies and in the world around us all the time. Without most of these processes, life would not be possible. *Living By Chemistry* supports teachers in fostering students' wonder and curiosity about the world around them.

Science as Guided Inquiry

Living By Chemistry is the product of a decade of research and development in high school classrooms, focusing on optimizing student understanding of chemical principles. The curriculum was developed with the belief that science is best learned through first-hand experience and discussion with peers. Guided inquiry allows students to actively participate in, and become adept at, scientific processes and communication. These skills are vital to a student's further success in science as well as beneficial to other future pursuits.

The *Living By Chemistry* curriculum provides you with a student-centered lesson for each day. Students have opportunities to ask questions, make scientific observations, collect evidence, and formulate scientific hypotheses and explanations. In each lesson, students discover concepts and communicate ideas with peers and with the teacher. Formal definitions and formulas are frequently introduced *after* students have explored, scrutinized, and developed a concept, providing more effective instruction.

Thinking Like a Scientist

Using guided inquiry as a teaching tool promotes scientific reasoning, critical-thinking skills, and a greater understanding of the concepts. Students develop their own logical conclusions and discover chemistry concepts for themselves, rather than accepting and memorizing facts. The ultimate goal is to foster students who think like scientists and understand the nature of scientific practice. Students learn to study the natural world by asking questions, and proposing explanations based on evidence. They learn to reflect on their ideas and review their work and that of their peers, and to effectively communicate scientific concepts they have discovered.

Chemistry for All Students

Chemistry is at the core of many aspects of our daily lives. Now, more than ever, the world needs citizens who can make informed decisions about their health, the environment, energy use, nutrition, and safety. In addition, chemistry is a required course for a myriad of different career paths relating to science, engineering, health, and the environment. The *Living By Chemistry* curriculum helps you to promote scientific literacy and support all students in developing valuable skills that extend well beyond the classroom.

Sequenced for Understanding

The sequencing of topics in *Living By Chemistry* is purposeful and well-tested. The topics are ordered and presented in a way that optimizes understanding. In addition, the curriculum covers all necessary standards and concepts. (See the Content Coverage Chart on pages xvi–xvii.) You may be tempted to reorder topics or front-load detailed information when a topic is introduced—this is not necessary because *Living By Chemistry* is a spiraling curriculum in which topics are revisited in increasing depth throughout the course. Our experience confirms that by building a solid foundation and by scaffolding all the topics—including the mathematics—*Living By Chemistry* can help you prepare your students for even the most challenging topics.

A Typical Day

Start With Student Understanding

Students come into any class with prior knowledge, assumptions, and misconceptions. When students make sense of new evidence and revise their thinking to accommodate it, they build true understanding. *Living By Chemistry* supports this process by allowing students to build their understanding based on experiences and including discussion questions specifically designed to challenge them to share their reasoning. Lessons are constructed to lead students to a more complete understanding.

The first step in this learning process is to discover students' current understanding of a concept or subject, through discussion or written questioning. Students' understanding is likely to be flawed or incomplete, so it is vital to create a safe atmosphere in the classroom for the sharing of these ideas. To do this, the *Living By Chemistry* curriculum uses a modified version of the 5Es model of teaching and learning. Originally created by a team led by Principal Investigator Rodger Bybee at the Biological Sciences Curriculum Study (BSCS), the 5Es describe different stages of a learning sequence: Engage, Explore, Explain, Elaborate, and Evaluate.

Lessons That Promote Understanding

Engage: At the start of each class, students are immediately engaged in a brief warm-up exercise, called a ChemCatalyst, that focuses on the main goal of the lesson. The purpose of the ChemCatalyst is to determine students' prior knowledge on the subject and encourage participation. In most cases, you can listen to student ideas and ask for explanations without judgment or correction.

> **ChemCatalyst**
>
> 1. If you were to drop a spoonful of salt, NaCl, into a glass of water, what would happen?
> 2. If you were to drop a gold ring into a glass of water, what would happen?
> 3. What do you think is different about the atoms of these two substances? Why do you suppose the gold atoms don't break apart?

Explore: For the next 15 to 20 minutes of class, students explore the key chemistry topics covered that day. Depending on the lesson type, they might solve problems, analyze data, perform a laboratory experiment, build models, or complete a card sort activity. They might also watch a brief demonstration or a computer simulation and try to provide explanations for their observations. This is a chance for students to think and build their own understanding, and, most importantly, to support their ideas with evidence. Generally, students work collaboratively in small groups or pairs. During this portion of the lesson, you can circulate from group to group, offering guidance, asking questions, and helping students to refine their ideas.

Explain and Elaborate: A teacher-led discussion follows the Explore portion of the class. It allows students to connect their conceptual understanding from the lesson with new chemistry concepts, ideas, tools, or definitions that make up the learning objectives. Sample discussion questions are provided for you, along with summaries of the key points to be covered.

Evaluate: At the end of class, a final Check-in question provides both you and students with a quick assessment of their grasp of the day's main concepts.

> **Check-in**
> Predict whether $MgSO_4(aq)$, commonly known as Epsom salts, will conduct electricity. State your reasoning.

Homework: Each lesson is accompanied by a reading assignment in the student textbook. Usually this amounts to about two or three pages of reading that reviews and reinforces what was learned in class, followed by exercises. The reading includes diagrams, photos, worked examples, and real-world connections. *Living By Chemistry* is designed so that each reading *follows* its corresponding lesson. This reinforces the concepts developed in class, and therefore optimizes the effectiveness of both the classroom experience and the reading component.

Other Resources

Kits

Specialty items specifically designed for *Living By Chemistry*, such as card decks and molecular modeling sets, are provided in kits. Kits contain enough materials for classes of up to 32 students. Replacements are available for purchase. Additional materials needed for teaching each unit, including lab equipment and chemical supplies, are listed online at **www.keypress.com/keyonline**.

Lesson Presentations

Lesson presentations are also available in Microsoft® PowerPoint® format. For these and other resources, please visit **www.keypress.com/keyonline**.

SciLinks

Our partnership with SciLinks allows you to access the best resources available on the Internet for specific topics. Links take you to websites selected by the National Science Teachers Association.

Tips and Classroom Management Strategies

Share the Approach

For some students, the *Living By Chemistry* approach will be new and different from their prior science classroom experiences. Let your students know that they will be taking an active role in their own learning, and that seeking the "right" answer will not always be the best strategy. Let them know that you will often be more interested in *why* they think something rather than whether or not what they think is scientifically correct. This will help to establish an open and safe atmosphere where students can share ideas and build their knowledge of chemistry.

Begin With the Basics

Living By Chemistry begins with the basics. We don't assume that students come into your classroom with any exposure to chemistry or its concepts. This scaffolded approach means that students new to chemistry are not at an instant disadvantage.

At the same time, students who are acquainted with chemistry vocabulary and principles have an opportunity to clarify and expand their understanding. You'll find that while lessons may start simply, they build to embrace sophisticated concepts and principles. Although *Living By Chemistry* might appear to start more slowly than other courses you've seen, all students benefit in the long run by developing a deeper and more comprehensive understanding.

Adjust Pacing to Match the Class

The flexibility of the *Living By Chemistry* curriculum allows you to adjust pacing to meet the needs of each class. For example, for an honors class the pacing and expectations can be increased. Some lessons can be completed as homework, more projects and Internet research can be assigned, and concepts can be pursued to more sophisticated levels. Other classes may work at a slower pace, with more time spent discussing, practicing, and clarifying concepts. For these classes, you may wish to occasionally spend an extra day on a lesson. For double-periods, two lessons can often be easily merged into one. (See Pacing Guides on page xx.)

Cooperative Learning

In the *Living By Chemistry* curriculum, students work cooperatively in groups, gaining expertise by articulating ideas and communicating concepts. This is one of the best ways to become proficient in a discipline. Students benefit from seeing a variety of approaches to a single problem, and they use their group as a sounding board for their ideas, rather than struggling alone or having to ask you.

You might assign new groups at the start of each unit. Some teachers select groups based on students' strengths or other criteria, while some teachers prefer to assign groups randomly.

One strategy that lends to effective group work is a fair division of tasks among the group members. Each member could be assigned a role, such as spokesperson, recorder, equipment and lab safety person, or facilitator. In this way, the work is shared, everyone is involved, and everyone has an opportunity to practice communication, leadership, and responsibility.

Success for More Students

Extensive field-testing and research over the past decade have shown that students of all ability levels perform better with *Living By Chemistry*. One of the gratifying comments that we hear frequently from teachers is that they find many more students participating and engaged, including those who were not participating previously. We hope your experience with this curriculum meets and exceeds your expectations, and we welcome your feedback. Contact us at editorial@keypress.com.

Features of the Teacher Guide

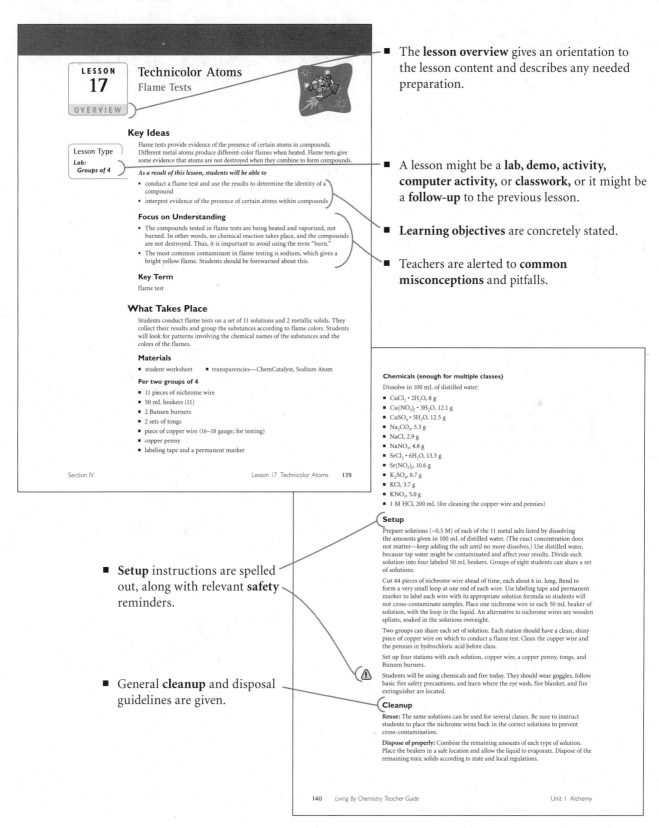

- The **lesson overview** gives an orientation to the lesson content and describes any needed preparation.

- A lesson might be a **lab, demo, activity, computer activity,** or **classwork,** or it might be a **follow-up** to the previous lesson.

- **Learning objectives** are concretely stated.

- Teachers are alerted to **common misconceptions** and pitfalls.

- **Setup** instructions are spelled out, along with relevant **safety** reminders.

- General **cleanup** and disposal guidelines are given.

xiii

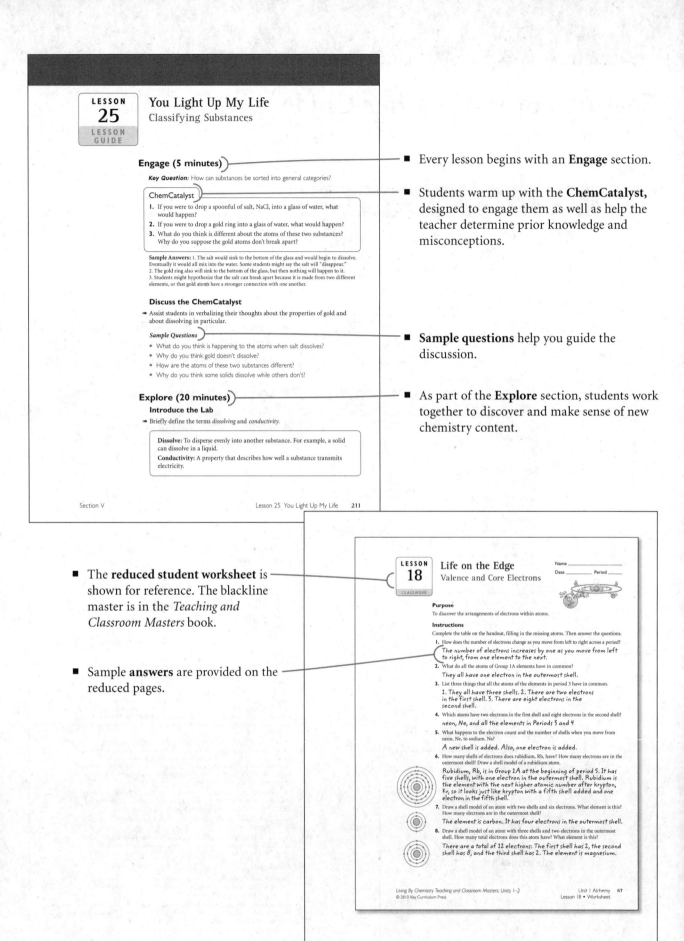

- Every lesson begins with an **Engage** section.

- Students warm up with the **ChemCatalyst**, designed to engage them as well as help the teacher determine prior knowledge and misconceptions.

- **Sample questions** help you guide the discussion.

- As part of the **Explore** section, students work together to discover and make sense of new chemistry content.

- The **reduced student worksheet** is shown for reference. The blackline master is in the *Teaching and Classroom Masters* book.

- Sample **answers** are provided on the reduced pages.

Explain and Elaborate (15 minutes)
Share Students' Generalizations about Conductivity
→ Summarize students' generalizations and write them on the board as students share them. Ask for consensus on the statements as you accept them.

Sample Questions
- What generalizations can you make about the substances that did not light up the bulb?
- What generalizations can you make about the substances that did light up the bulb?
- Based on your data, why do you think the sports drink lit up the bulb when dissolved?

Key Points
Generalizations about substances that *do not* light up the bulb:
- Compounds made up of C, H, and O atoms do not conduct electricity.
- Compounds made up entirely of nonmetals do not light up the bulb.
- Compounds made up of a combination of metals and nonmetals do not light up the bulb when they are in their solid form.

Generalizations about substances that *do* light up the bulb:
- Everything that lights up the bulb has a metal atom in it.
- Compounds made of metal and nonmetal atoms, such as salts, light up the bulb when they are dissolved in water. (The sports drink is a solution of water, various salts, sugar, and a dye.)
- Metal solids light up the bulb.

Analyze the Results
(T) → Use the transparency Solubility and Conductivity to summarize the results. Ask students to assist you with filling in the substances at the bottom of the chart.
→ After sorting the substances, introduce the terms *soluble* and *insoluble*. Label the appropriate boxes with these terms.

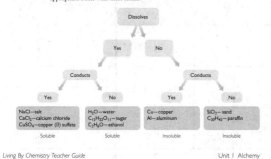

Sample Questions
- What statement can you make about ionic compounds and dissolving?
- What generalizations can you make about substances made up entirely of metal atoms?
- What generalizations can you make about substances made up entirely of nonmetal atoms?

Key Point
We can place all the substances tested into one of the four categories. A substance that dissolves in another substance is said to be "soluble" in that substance. A substance that does not dissolve in another substance is said to be "insoluble" in that substance. Once we find out whether a substance dissolves in water, we can further sort according to whether the dissolved substance conducts electricity. In the next lesson we will take a closer look at these categories in order to figure out what else the substances have in common, or what might account for their common properties.

> **Soluble:** Describes a substance that is capable of being dissolved in another substance.
> **Insoluble:** Describes a substance that is incapable of being dissolved in another substance.

Notice that substances that conduct electricity are either solid metals or ionic compounds dissolved in water. Substances made entirely of nonmetal atoms, such as sugar, do not conduct electricity.

Note: Some substances are considered slightly soluble. More discussion of solubility will take place in Unit 4: Toxins.

Wrap-up
Key Question: How can substances be sorted into general categories?
- Not all substances dissolve in water.
- Not all substances conduct electricity.
- Solid metals and metal–nonmetal compounds dissolved in water conduct electricity.

Evaluate (5 minutes)
Check-in
Predict whether $MgSO_4(aq)$, commonly known as Epsom salts, will conduct electricity. State your reasoning.

Answer: $MgSO_4$ dissolves in water and it contains both metal and nonmetal atoms so it will conduct electricity.

Homework
Assign the reading and exercises for Alchemy Lesson 25 in the student text.

Content Coverage Chart

Concepts are introduced and then reinforced, often across different units. Coverage usually consists of an introduction (I), then practice (P), and finally teaching to mastery (M).

	Unit and Section								
	Alchemy					Smells			
	I	II	III	IV	V	I	II	III	IV
Content									
Atomic and Molecular Structure			I	P	P	P	P	P	M
Chemical Bonds					I	P	P	P	P
Conservation of Mass and Stoichiometry	I	P	P						
Gases and Their Properties									
Acids and Bases		I				P			
Solutions		I							
Energy/Chemical Thermodynamics									
Reaction Rates						I			
Chemical Equilibrium									
Organic Chemistry and Biochemistry						I	P	P	P
Nuclear Processes			I/P						
Investigation and Experimentation	I	P	P	P	P	P	P	P	P

Unit and Section													
Weather			Toxins					Fire				Showtime	
I	II	III	I	II	III	IV	V	I	II	III	IV	I	II
										M			
			P	P	P	P	P	P	P	P	M		
I	P	M											
						P						P	M
					P	P	P				P	P	M
	I							P	P	P	P	M	
										P		P	P
												I	P
P	P	P	P	P	P	P	P	P	P	P	P	P	M

Introduction to Smells: Molecular Structure and Properties

Smell as Context

Smell is a highly intriguing and familiar context that activates your students' curiosity. There's plenty of room in this unit to share information, theories, memories, and experiences. The students get to be the experts. As the teacher, you can choose how long or involved the discussions become. The goal is to spark an interest that will be sustained throughout the unit.

Students become more actively involved in learning when they can relate to the subject matter taught. When the immediate relevance of chemistry concepts is demonstrated, students find the chemistry easier to grasp and apply.

Smell Classification as a Field of Research

The mechanism by which we detect the variety of odors in our environment is not yet completely understood. For decades, scientists have been trying to identify what they consider to be the "primary odors" and to agree upon a naming system for these smells. Most studies provide evidence that scents can be categorized into more or less discrete classes.

For this unit, we have chosen to focus on five terms that have been consistently chosen by researchers to describe groups of common odors: minty, fishy, sweet, putrid, and camphor. The category we call "sweet" includes both fruity and floral smells for convenience's sake. While not every odor observed in daily life will fit into these five categories, they do account for a large number of the smells one might encounter. These groups provide us with a common language and a structure to use in our pursuit of smell chemistry. For a summary of smell classifications covered in this unit and the chemistry behind them, see page 159.

Content Driven by Context

This unit establishes students' understanding of molecules through tangible experience. Students can draw on their own experience with scents to analyze molecular properties and behavior. They begin by examining smells and molecular data for several compounds. Students are then asked to formulate a hypothesis based on patterns in the data that relates smell and molecular structure. The lessons are designed so that students continue to test their hypotheses, refining their conclusions as new evidence is discovered. Throughout the unit, students learn the chemistry related to smell. They explore molecular formulas, Lewis dot symbols, structural formulas, polarity, molecular size and shape, and bonding patterns. They learn to identify particular functional groups within molecules and are introduced to organic chemistry and the chemistry of living things.

Variations in Experience

Somewhere between 0.5% and 1% of the population have a condition called anosmia, and cannot perceive smell. Students with anosmia or respiratory difficulties can allow others to smell for them during the smelling activities. There are also variations in the ability to smell. These inconsistencies in smell perception may come up in the classroom. These discrepant events represent a more specialized area of inquiry that students can pursue in the project at the end of the unit.

However the majority experience will guide this particular unit. We are trying to teach students the *basics* of smell chemistry. Encourage those with dissenting opinions to continue to speak up, but rely on consensus to direct the activities.

The Use of Models to Explain Molecules

Citronellol, $C_{10}H_{20}O$

This unit continues to use models to explain matter. Students explore 2-dimensional structural formulas and 3-dimensional ball-and-stick and space-filling models. They learn to make predictions about the smell of molecules based on evidence that they discover using models. Current theories propose that molecules of different shapes fit into matching receptor sites located on the surface of tiny hair-like cilia, which line the nasal passages. Once the molecule has docked, it triggers a neural impulse to the portion of the brain responsible for smell. Students explore this theory with models and they learn to classify and categorize different compounds.

Building Understanding

In Alchemy, students began their study of matter and atomic properties. In this unit, students gradually construct understanding about molecules and their properties as they explore ideas about smell. This foundation in molecular structure and behavior serves to prepare students for a better understanding of the gas laws and chemical reactions, the subjects of later units.

Section I is an introduction to the smells context. Students first compare smell categories and consider the question, "What is the connection between smell and molecular composition?" They develop a hypothesis which they have the opportunity to revise as they explore Lewis dot structures and octet rule and functional groups. The section culminates with a lab in which they synthesize an ester from an alcohol and an organic acid.

In **Section II,** students work with ball-and-stick models to study the role of electron pairs and molecular shape. They rethink and refine the hypotheses they proposed in the first section as new smell compounds are introduced. They continue their study using space-filling models to generate links between molecular shape and smell. In this section, they consider the receptor site theory for smell.

Section III focuses on interactions between molecules. Students observe evidence of polarity in the lab and learn about electronegativity. Electronegativity values are used to determine the polarity of molecules and the direction of bond dipoles. In this section, students answer the question, "How are polarity, phase, shape and bonding patterns related to smell?"

Section IV focuses on biological molecules. Students study the "handedness" of molecules and are challenged to distinguish between mirror image isomers. In this section, students are introduced to amino acids and build models to see how amino acids link and fold to form proteins. Students use this information as they revisit receptor site theory and review the chemistry related to smell.

Pacing Guides

Standard Schedule

Day	Suggested Plan
1	Lesson 1
2	Lesson 2
3	Lesson 3
4	Lesson 4
5	Lesson 5
6	Lesson 6
7	Lesson 7
8	Lesson 8, Section I Review
9	Section I Quiz
10	Lesson 9
11	Lesson 10
12	Lesson 11
13	Lesson 12
14	Lesson 13
15	Lesson 14, Section II Review
16	Section II Quiz
17	Lesson 15
18	Lesson 16
19	Lesson 17
20	Lesson 18
21	Lesson 19, Section III Review
22	Section III Quiz
23	Lesson 20
24	Lesson 21, Section IV Review
25	Section IV Quiz, Lesson 22
26	Unit Review
27	Unit Exam
28	Lab Exam (optional)

Block Schedule

Day	Suggested Plan
1	Lessons 1 and 2
2	Lessons 3 and 4
3	Lessons 5 and 6
4	Lesson 7 and 8, Section I Review
5	Section I Quiz, Lesson 9 and 10
6	Lessons 11 and 12
7	Lessons 13 and 14, Section II Review
8	Section II Quiz, Lesson 15
9	Lessons 16 and 17
10	Lessons 18 and 19, Section III Review
11	Section III Quiz, Lesson 20
12	Lessons 21, Section IV Review
13	Section IV Quiz, Lesson 22
14	Unit Review
15	Unit 2 Exam, Lab Exam (optional)

SECTION I

Speaking of Molecules

Section I introduces students to the smell context. In Lesson 1, students sample several smells and examine molecular data for these compounds. They discover patterns in the data that relate smell to molecular composition. Lessons 2 and 3 focus on structural formulas and the covalent bonding tendencies of hydrogen, oxygen, nitrogen, and carbon. Lessons 4 and 5 introduce students to Lewis dot symbols and the octet rule. In Lesson 6, students explore the relationship between functional groups and smell. As students explore and integrate the topics in this section, they refine their hypotheses about how smell is related to a molecule's composition and structure. Finally, in Lessons 7 and 8, students participate in a lab to synthesize a sweet-smelling ester from an alcohol and an organic acid and analyze the results.

In this section, students will learn

- to interpret molecular formulas
- to create molecular structures from molecular formulas
- to use Lewis dot symbols to predict molecular bonding and structure
- to identify and name functional groups within molecules
- about relationships among molecular formula, molecular structure, functional group, and smell

LESSON 1	Sniffing Around
OVERVIEW	Molecular Formulas

Lesson Type
Activity:
Groups of 4

Key Ideas

We can talk about molecules by using molecular formulas and chemical names. A preliminary look at the connection between smell and chemistry shows relationships among the molecular formula, chemical name, and smell of a compound.

As a result of this lesson, students will be able to

- detect patterns in chemical formulas and relate these patterns to a molecular property
- create a hypothesis based on analysis of data

Focus on Understanding

- Students with very strong allergies or asthma should allow others to do the smelling for them and then work from their classmates' data.
- In rare instances, a student might have no sense of smell or some students' sense of smell will differ from the norm. These cases can make a good discussion topic.

Key Term

molecular formula

What Takes Place

Working in groups, students sample the smells of five substances. Then they examine data sheets that give the molecular formula and chemical name of each of the five compounds. Students work in groups to look for patterns in the data. These patterns are discussed and summarized on the board. The class forms a hypothesis about how the smell of a compound can be predicted from certain molecular information. This hypothesis will be expanded and revised several times during the unit as students consider more information.

Materials

- student worksheet
- 5 plastic pipettes

Per group of 4 (may be re-used for multiple classes)

- vials A–E
- 5 cotton balls
- spearmint oil, 5 mL

- pineapple extract, 5 mL
- banana extract, 5 mL
- peppermint oil, 5 mL
- fish oil, 5 mL

Setup

Prepare sets of vials A–E as described in the table. Label the vials only with the letters A–E. Place a cotton ball in each vial, then use the plastic pipettes to place three to five drops of each essence in the appropriately lettered vial. Use a different pipette for each essence. Place sets of the five vials in reusable plastic bags to make it easy to distribute them to groups of students.

Label	Contents	Smell
A	spearmint oil	minty
B	fish oil	fishy
C	pineapple extract	sweet
D	banana extract	sweet
E	peppermint oil	minty

Cleanup

Save the vials for reuse in other classes and over the course of this unit. Within a few weeks, however, you will need to remove the cotton balls and air out the vials. Otherwise, the different smells will mix with one another and begin to smell putrid.

When you are finished using vials A–E in all your classes, remove the cotton balls from the vials, place them in a plastic bag, and dispose of them. Let the vials air out in a hood or rinse them with acetone for reuse the next time you do this unit.

LESSON 1 GUIDE

Sniffing Around
Molecular Formulas

Engage (5 minutes)

Key Question: What does chemistry have to do with smell?

ChemCatalyst

1. What do you think is happening when you smell something?
2. Why do you think we have a sense of smell?

Sample Answers: 1. Students might say the nose is detecting a "gas" or different "molecules." Ask them to clarify and to give evidence. Keep the discussion open-ended. 2. Students might say the sense of smell is to help us find food, detect danger (skunk, toxins), enjoy life (flowers), find a mate, and so on.

Discuss the ChemCatalyst

➧ Stimulate interest in the context and get students to think about how the act of smelling happens.

Sample Questions

- Why do some things smell good and other things smell bad?
- How does a smell get from one place to another?
- Describe what you think your nose is detecting when you smell something.
- What do you think chemistry has to do with smell?

Explore (15 minutes)

Introduce the Activity

➧ Review the concept of a molecule and introduce molecular formulas.

- Tell students they will examine molecular covalent substances in this unit. These substances consist of molecules. Each molecule stays together as a unit and is represented by a molecular formula.

Molecular formula: The chemical formula of a molecular substance, showing the types of atoms in each molecule and the ratios of those atoms to one another.

⇒ Instruct students on safe smelling techniques.
- Chemicals may have very strong odors or be caustic. Remind students to use a *wafting* technique when they smell anything in the laboratory: They should use their hand to draw air toward them, not sniff directly from the container.
- The essences are food grade, and the smells are no stronger than those we encounter daily. However, any students with asthma or allergies should not do the smelling part of the activity.

⇒ Pass out vials A–E. Students work in groups of four.
- Emphasize to students that they need to put the caps back on the vials immediately after smelling and not exchange caps among the vials.
- Before collecting the vials, tell students to tighten all the caps.

Guide the Activity

⇒ Encourage students to use the convention shown in the table when writing their molecular formulas: carbon, then hydrogen, then oxygen (e.g. C_3H_8O).

LESSON 1 ACTIVITY

Sniffing Around
Molecular Formulas

Name _____
Date _____ Period _____

Purpose

To explore the connection between chemistry and smell.

Materials

- vials A–E

Part 1: Smelling

Smell the five mystery smells in vials A–E using the wafting technique. Replace the caps immediately after smelling and take care not to mix them up.

1. Identify the smells yourself as either fishy, minty, or sweet. Then discuss the smells as a group and reach consensus on the smell classification.

Vial	Your classification	Group consensus
A	Answers may vary.	minty
B		fishy
C		sweet
D		sweet
E		minty

Part 2: Looking for Patterns

1. Enter the group consensus smells in the smell data table.

Smell Data

Vial	Smell	Chemical name	Molecular formula
A	minty	L-carvone	$C_{10}H_{14}O$
B	fishy	phenylethylamine	$C_8H_{11}N$
C	sweet	pentyl propionate	$C_8H_{16}O_2$
D	sweet	isopentyl acetate	$C_7H_{14}O_2$
E	minty	menthone	$C_{10}H_{18}O$

2. Look for patterns in the data. Write down at least eight patterns you discover between your data and the various smells.

Patterns: Answers will vary. Some examples are provided.
1. *All the molecules have carbon and hydrogen atoms.*
2. *Molecules that smell fishy have a nitrogen atom.*
3. *Molecules that end with "-ate" smell sweet.*
4. *Minty-smelling substances have ten carbon atoms.*
5. *Things that smell similar have similar endings to their names.*
6. *Substances with oxygen have even numbers of hydrogen atoms.*
7. *Molecules that end in "-one" smell minty.*
8. *Substances that smell sweet have two oxygen atoms.*

Questions

1. Why do you think there are sometimes disagreements over how to classify smells?

   Some people might not be able to smell certain substances. Describing smells is a subjective experience; that is, it is an opinion with no right or wrong answer.

2. From the data, what generalization could you make about substances that contain oxygen atoms?

   Substances that contain oxygen atoms smell sweet or minty. They don't smell fishy.

3. Which patterns might be useful in helping you predict smells?

   Answers will vary. Above, numbers 2, 3, 4, 5, 7, and 8 are potentially helpful in predicting smell. Numbers 1 and 6 are not.

4. **Making Sense** What evidence is there that smell, molecular formula, and chemical name are related?

   Many of the patterns suggest a connection between molecular formula, name, and smell. For example, both molecules with one oxygen atom smell minty and the names both end in "-one."

Explain and Elaborate (20 minutes)

Assist Students in Sharing Their Experience of Sampling Smells

➡ Draw this table on the board. Fill in as groups share their smell classifications.

Vial A	Vial B	Vial C	Vial D	Vial E
minty	fishy	sweet	sweet	minty

Sample Questions

- Which smell classification did you use for each of the five mystery smells?
- Which vials would you put in the same category? (A and E, C and D)
- Why do you think there are sometimes disagreements over how to classify the smells of different substances?

Collect a List of Patterns Generated by the Students

➡ Write the patterns on the board as the students share them.

Sample Questions

- What patterns did you discover?
- According to the data, if a molecule has one oxygen atom, how would you expect it to smell? (minty) What if it has a nitrogen atom? (fishy) What if it has two oxygen atoms? (sweet)
- Which other patterns help you predict the smell of a molecule?

Patterns students might notice

All the molecules:

- All the molecules have H and C atoms.
- There are more H atoms than C atoms.
- The names of molecules that have similar smells have similar endings.
- Molecules that smell good all have even numbers of H atoms.
- Smell seems to be related to the atoms other than C and H atoms.

Sweet-smelling molecules:

- Molecules that smell sweet have two O atoms.
- Molecules that end with "-ate" smell sweet.

Minty-smelling molecules:

- Molecules that smell minty have one O atom and no N atoms.
- Minty-smelling substances have ten C atoms and end in "-one."

Fishy-smelling molecules:

- Molecules that smell fishy have one N atom.
- The molecules with names that end with "-ine" smell fishy.

Note: While many of these patterns are useful in predicting smells, there will be exceptions, and some statements need refinement. The goal is to refine these ideas so that we improve our accuracy in predicting smells.

Come Up with a Hypothesis About Predicting Smells

➤ At the appropriate point in the discussion, write a hypothesis about predicting smells on the board. Allow students to refine it if they wish.

Sample Questions

- If I gave you a new smell vial, what would you want to know about the substance inside in order to predict its smell?
- Do you think you can predict the smell of a substance simply from knowing its chemical name and molecular formula? Why or why not?
- What should we do to test this hypothesis?

Key Point

A possible hypothesis is "The smell of a substance can be predicted if you know its name and/or its chemical formula." For instance, molecules with two oxygen atoms may all smell sweet. Molecules containing nitrogen may all smell fishy. Molecules containing one oxygen atom may all smell minty. Molecules ending in "-ate" may all smell sweet. And so on. As students learn more about molecules and smell properties throughout the unit, they will have a chance to refine their hypotheses.

Wrap-up

Key Question: What does chemistry have to do with smell?

- Smell appears to be related to molecular formula and chemical name.

Evaluate (5 minutes)

> Check-in
>
> 1. How would you expect a compound with the molecular formula $C_8H_{16}O_2$ to smell? Explain.
> 2. How sure are you of your prediction?

Answers: 1. From the preliminary data, you can predict it will smell sweet because it has two oxygen atoms. 2. You cannot be all that sure, because you have smelled only five molecules. You need to collect more data.

Homework

Assign the reading and exercises for Smells Lesson 1 in the student text.

LESSON 2 OVERVIEW

Molecules in Two Dimensions
Structural Formulas

Key Ideas

Structural formulas are models that chemists use to show how the atoms in a molecule are connected to one another. Two molecules might have the same molecular formula but different structures. Such molecules are called isomers.

As a result of this lesson, students will be able to

- describe the difference between structural formulas and molecular formulas
- recognize isomers

Focus on Understanding

- It might take some time for students to be able to visually distinguish the details of structural formulas and to realize that the same molecule can be represented in different ways on the page.
- Functional groups will be introduced in a later lesson. For now students only need to understand that atoms can be connected in different ways to form different overall structures.

Key Terms

structural formula
isomer

What Takes Place

At the beginning of class, students predict how three new mystery substances will smell based on their molecular formulas and chemical names. Students then test their predictions by smelling these new substances. They examine the structural formulas of these same three molecules, looking for similarities and differences. Students explore the orientation of structural formulas and look at ball-and-stick models to become more proficient at differentiating between different molecules and different orientations. Students are introduced to isomers.

Materials (for a single class or for multiple sections)

- student worksheet
- transparency—Structural Formulas
- molecular modeling set
- 3 pipettes
- apricot perfume oil, 5 mL
- butyric acid, 5 mL
- rum flavor extract, 5 mL

> **Lesson Type**
> Activity:
> Groups of 4

Per group of 4

- vials F–H
- 3 cotton balls

Setup

Prepare sets of vials F–H according to the table. Label the vials with the letter only. Place a cotton ball in each vial, then use the plastic pipettes to deliver three to five drops of each stock smell solution to the appropriately lettered vial. Use a new pipette for each essence. You should prepare the vials in a hood or outdoors, especially vial G.

Label	Contents	Smell
F	apricot perfume oil (ethyl pentanoate)	sweet
G	butyric acid	putrid
H	rum flavor extract (ethyl acetate)	sweet

Build one ball-and-stick molecular model each of butyric acid, ethyl acetate, and 1-propanol (molecule 1 from the worksheet). Here are the structural formulas and the pieces you will need. Use 4-hole carbons.

$$\begin{array}{c} \text{H H H O} \\ \text{H—C—C—C—C—O—H} \\ \text{H H H} \end{array} \quad \begin{array}{c} \text{H O H H} \\ \text{H—C—C—O—C—C—H} \\ \text{H H H} \end{array} \quad \begin{array}{c} \text{H H H} \\ \text{H—C—C—C—O—H} \\ \text{H H H} \end{array}$$

Butyric acid   Ethyl acetate   1-propanol

	Carbons (black)	Oxygens (red)	Hydrogens (white)	Single bonds (straight sticks)	Double bonds (curved sticks)
butyric acid	4	2	8	12	2
ethyl acetate	4	2	8	12	2
1-propanol	3	1	8	11	0

Cleanup

Save the vials for reuse in other classes and over the course of this unit. Within a few weeks, however, you will need to remove the cotton balls and air out the vials. Otherwise, the substances' smells will mix with one another and begin to smell putrid.

When you are finished using vials F–H in all your classes, remove the cotton balls from the vials, place them in a plastic bag, and dispose of them. Let the vials air out in a hood or rinse them with acetone for reuse the next time you do this unit.

LESSON 2 GUIDE

Molecules in Two Dimensions
Structural Formulas

Engage (5 minutes)

Key Question: How can molecules with the same molecular formula be different?

ChemCatalyst

Predict the smells of these three new molecules. Provide evidence to support your prediction.

Vial F: ethyl pentanoate $C_7H_{14}O_2$
Vial G: butyric acid $C_4H_8O_2$
Vial H: ethyl acetate $C_4H_8O_2$

Sample Answer: Some students might predict that all three vials will smell sweet, based on their chemical formulas and the presence of two oxygen atoms. However, vial G may smell different, based on the chemical name.

Discuss the ChemCatalyst

Sample Questions

- What smells did you predict for each vial? What was your reasoning?
- What do these molecules have in common? How do they differ?
- Do you think vial G and vial H will smell identical? Why or why not?
- What information do you need in order to find out if your predictions are correct?

Explore (15 minutes)

Guide the Activity

⟹ Pass out worksheets. Students can work individually.

⟹ Ask students to write their predictions in the table before you pass out the vials with the new molecules.

⟹ Pass out vials F–H to groups of four students.

⟹ After smell sampling is complete, ask students to tighten the caps on the vials. Collect the vials.

⟹ When students get to question 6, hold up the model of 1-propanol. Show them how different spheres represent different kinds of atoms while the "sticks" represent bonds. Show them how turning the model or rotating the bond can make it look different, as in the third or fifth drawing, but does not change the structure of the molecule (i.e., it was not taken apart and put back together).

LESSON 2 ACTIVITY
Molecules in Two Dimensions
Structural Formulas

Name _____
Date _____ Period _____

Purpose
To compare the structures of molecules.

Materials
- vials F–H

Part 1: Test Your Predictions
Write your predictions in the table. Then carefully smell vials F, G, and H.

Vial	Chemical name	Molecular formula	Predicted smell	Actual smell
F	ethyl pentanoate	$C_7H_{14}O_2$	sweet	sweet
G	butyric acid	$C_4H_8O_2$	(Answers vary)	putrid
H	ethyl acetate	$C_4H_8O_2$	sweet	sweet

1. What name would you give to the smell category that vial G might belong in?

   Possible answers: putrid, nasty, rotten

2. What could account for two molecules with the same molecular formula having different smells?

   Answers will vary. There must be some other difference between these molecules, perhaps the way the atoms are connected.

Part 2: Examine the Structures
Below are structural formulas of each of the three substances. They show how the atoms in each molecule are connected.

Molecule F   Molecule G   Molecule H

Questions

1. List three similarities between molecules G and H.

   Both have the same numbers and types of atoms. Both have two oxygen atoms, a carbon atom that is connected to an oxygen atom with two lines, and a carbon atom that is connected to an oxygen atom with one line.

2. List two differences between molecules G and H.

   Possible answers: An oxygen atom is in the middle of molecule H but near the end of molecule G. Molecule G has four carbon atoms in a row, but molecule H has only two carbon atoms in a row.

3. List three similarities between the two sweet-smelling molecules.

   Possible answers: Molecules F and H both have two carbon atoms connected to five hydrogen atoms on one end. Both have an oxygen atom between two carbon atoms and a carbon atom connected by two lines to an oxygen atom.

4. What do you suppose the lines in these drawings represent?

   Possible answer: The lines represent connections, or bonds.

5. From the evidence you have seen so far, how would you explain the differences in smell between molecules G and H?

   The atoms are connected differently, particularly the oxygen atoms, which results in different properties.

6. Five more structural formulas are shown here. Write their molecular formulas.

   Molecule 1: C_3H_8O
   Molecule 2: C_3H_8O
   Molecule 3: C_3H_8O
   Molecule 4: C_3H_8O
   Molecule 5: C_3H_8O

7. Molecules 1, 3, and 5 smell exactly the same. They represent the same molecule. Explain why.

   The atoms are all connected the same. The molecules are simply rotated or turned.

8. Molecules 1, 2, and 4 have different smells. Explain why.

   They have different structures. The oxygen atom is connected differently in the three molecules. They are not the same molecules.

9. **Making Sense** What evidence is there that the structure of a molecule is related to how it smells?

   Molecules G and H have the same molecular formula, but different structures and smells. Molecules F and H have different molecular formulas, but similar structural feature and similar smells.

10. **If You Finish Early** Draw molecule 4 so that it looks different on paper but still represents the same molecule.

Explain and Elaborate (20 minutes)
Discuss the Three New Smells—F, G, and H

➠ You might want to redisplay the ChemCatalyst.

Sample Questions

- How did your experience with the vials match your predictions?
- How do you explain your results?
- How did vial G smell? What smell classification might you use for vial G and other similar smells?

Key Point

Even though the molecules in vials G and H have identical molecular formulas, they have different smells and therefore must be different somehow. Smell chemists call substances that smell like vial G putrid. Even though the molecules in vial G have two oxygen atoms, like the sweet-smelling molecules, vial G smells disgusting. We will continue to use the classification "putrid" in the remainder of the unit. We now have four smell classifications: sweet, minty, fishy, and putrid.

➠ Show the model of 1-propanol, C_3H_8O, from the worksheet. Show how you can rotate the model or individual bonds to match the third and fifth drawings but not the second or fourth.

➠ Show the models of butyric acid, $C_4H_8O_2$ and ethyl acetate, $C_4H_8O_2$. Show that butyric acid can't be turned into ethyl acetate without taking the molecule apart.

Sample Questions

- What information is contained in a structural formula?
- How is the information in a structural formula different from the information in a molecular formula?
- When a molecular structure is different, does that mean the molecule is different? Explain.

Key Points

A structural formula is a two-dimensional drawing of a molecule showing how the atoms in a molecule are connected. Each line represents a covalent bond. A double line represents a double bond. Thus, every molecule in a particular substance has the same very specific structure that does not change.

> **Structural formula:** A drawing or diagram that a chemist uses to show how the atoms in a molecule are connected. Each line represents a covalent bond.

There are several ways to draw the same structural formula without changing the identity of the molecule. A molecule can be drawn forward, backward, or vertically, and it will still be the same molecule. This is because flipping the molecule in space, or turning it, does not change the way the atoms are connected. It is still the same substance.

When two molecules have the same molecular formula but different structural formulas, they are called isomers of each other. Isomers have distinct properties. Therefore, we would expect molecules with different structures to have different smells.

> **Isomers:** Molecules with the same molecular formula but different structural formulas.

$$\begin{array}{c} \text{H H H O} \\ \text{H—C—C—C—C—O—H} \\ \text{H H H} \end{array} \qquad \begin{array}{c} \text{H O H H} \\ \text{H—C—C—O—C—C—H} \\ \text{H H H} \end{array}$$

Butyric acid　　　　　　Ethyl acetate

Explore the Relationship Between Structural Formula and Smell

⇒ Display the transparency Structural Formulas.

Sample Questions

- What do the structural formulas tell you about the smells of molecules G and H? How are these two molecules different?
- Do you think you can predict the smell of a compound from only its chemical name and molecular formula? Why or why not?
- If you were given another putrid-smelling compound, what would you expect it to look like structurally?
- If you were given a new compound to smell, what information would you want to have in order to predict its smell?

Key Points

Molecules can smell different even if they have the same molecular formula. There is more to predicting smell than simply looking at the molecular formula. The structural formulas of molecules G and H show us that two substances can have identical molecular formulas but different structures. In particular, one of the oxygen atoms in molecule G is connected to a carbon atom and a hydrogen atom, whereas one of the oxygen atoms in molecule H is connected to two carbon atoms.

Molecules can smell similar even if they have different molecular formulas. Molecules F and H both smell sweet, even though their molecular formulas are different. However, both have a similar structural feature in the middle of the molecules: One oxygen atom is connected by two lines to a carbon atom, and another oxygen atom is between two carbon atoms.

Wrap-up

Key Question: How can molecules with the same molecular formula be different?

- Structural formulas show how the atoms in a molecule are connected.
- A molecular formula can be associated with more than one structural formula.

- Isomers are molecules with identical molecular formulas but different structural formulas.
- The smell of a molecule is a property that appears to be related to its structure.

Evaluate (5 minutes)

> **Check-in**
>
> For each compound, predict the smell or describe what information you would want in order to predict the smell.
>
> **a.** $C_6H_{12}O_2$ **b.** $C_6H_{15}N$

Answers: a. Either sweet or putrid. Having the chemical name of the molecule would help: "-ate" indicates a sweet smell; "acid" indicates a putrid smell. The structural formula would help too, because you could look for structural features. b. Fishy. We might want to smell a few more compounds containing nitrogen atoms to be sure they all smell fishy.

Homework

Assign the reading and exercises for Smells Lesson 2 in the student text.

LESSON 3 OVERVIEW

HONC if You Like Molecules
Bonding Tendencies

Key Ideas

Lesson Type
Classwork: Pairs

Within most molecules, hydrogen forms one bond, oxygen forms two bonds, nitrogen forms three bonds, and carbon forms four bonds. These bonding tendencies can be summarized with the mnemonic device HONC 1234.

As a result of this lesson, students will be able to

- create accurate structural formulas from molecular formulas
- identify and differentiate between isomers and molecules oriented differently in space
- explain and utilize the HONC 1234 rule

Focus on Understanding

- The orientation of a molecule in space might continue to be a source of confusion for students, making the accurate identification of isomers challenging.

Key Term

HONC 1234 rule

What Takes Place

Students are introduced to the HONC 1234 rule and use it to construct accurate molecular structures. Students work from molecular formulas to make two or three isomers for each formula. The sometimes subtle differences between different structural orientations versus different isomers are reinforced in this activity.

Materials

- student worksheet
- transparencies—ChemCatalyst and Check-in
- molecular modeling set

Setup

Build models of one or two C_3H_9N isomers from Question 3 on the worksheet for the Explain and Elaborate discussion.

LESSON 3 GUIDE

HONC if You Like Molecules
Bonding Tendencies

Engage (5 minutes)

Key Question: What are the rules for drawing structural formulas?

ChemCatalyst

Examine these molecules. What patterns do you see in the bonding of atoms of hydrogen, oxygen, carbon, and nitrogen?

Molecule K
Diisobutylamine; fishy

Molecule E
Menthone; minty

Sample Answer: Students will notice that hydrogen atoms bond only once. Carbon always has four bonds connecting it to other atoms. Oxygen has two bonds, and nitrogen has three.

Discuss the ChemCatalyst

➡ Discuss how H, O, N, and C atoms are connected in the structural formulas.

Sample Questions

- What patterns do you notice in the way atoms are connected?
- What do you think HONC 1234 means?
- How many connections do atoms of each element make with other atoms?

Explore (20 minutes)

Introduce the Activity

➡ Introduce the HONC 1234 rule. Write the letters H, O, N, and C on the board in a vertical column. Fill in the number of bonds as students respond.

hydrogen	H	1	(Hydrogen atoms form one bond.)
oxygen	O	2	(Oxygen atoms form two bonds.)
nitrogen	N	3	(Nitrogen atoms form three bonds.)
carbon	C	4	(Carbon atoms form four bonds.)

➡ Students can work individually or in pairs. Tell students they will use the HONC 1234 rule to construct structural formulas from molecular formulas.

LESSON 3 CLASSWORK

HONC if You Like Molecules
Bonding Tendencies

Name _____
Date _____ Period _____

Purpose

To practice constructing structural formulas from molecular formulas.

Procedure

Use the HONC 1234 rule and the general instructions below to create correct structural formulas from molecular formulas.

Example: C_3H_8O

Start by connecting the carbon atoms.   C—C—C

Next insert the nitrogen or oxygen atoms, either on the ends or somewhere in the middle of the carbon chain.

C—C—C—O   C—O—C—C   C—C(=O)—C

Add the hydrogen atoms last.

(three structural formulas with H atoms added)

Check that each atom follows the HONC 1234 rule.

Questions

1. Use the HONC 1234 rule to construct a structural formula for C_3H_8.

   H₃C—CH₂—CH₃ (structural formula shown)

2. Use the HONC 1234 rule to create two unique structural formulas for C_2H_6O.

   H₃C—O—CH₃   and   H—O—CH₂—CH₃

3. Use the HONC 1234 rule to create two unique structural formulas for C_3H_9N.

   Possible answers (any two):

   (four structural formulas shown)

4. There are at least two molecules with the molecular formula $C_2H_4O_2$. One is shown. Draw the other one. *Hint:* Each molecule has a double bond between a carbon atom and one of the oxygen atoms, C=O.

Molecule 1: H-C(H)(H)-C(=O)-O-H
Molecule 2: H-C(H)(H)-O-C(=O)-H

5. One of the molecules in Question 4 is sweet smelling, and the other is putrid. Predict which is which. Explain your reasoning.

The molecule that has the oxygen in the middle (molecule 2) will be sweet smelling based on other molecules encountered so far.

6. **Making Sense** These two molecular structures are incorrect according to the HONC 1234 rule. What specifically is wrong with each? Correct them by drawing new structures.

Molecule 1: C(H)(H)(H)-O-C(H)(H)(H) with extra bonds
Molecule 2: C(H)-H-O-H-C(H)

(Corrections will vary)

In molecule 1, the oxygen has four bonds instead of two and the carbon bonds have three bonds instead of four; remove two hydrogen atoms and place one each on the carbon atoms. In molecule 2, two of the hydrogen atoms have two bonds instead of one and the two carbon atoms have only three bonds.

7. **If You Finish Early** Try to draw a third structural formula for the molecular formula in Question 3.

See the answers for Question 3.

Explain and Elaborate (15 minutes)

Process the Different Structures from the Classwork

➡ Ask students to come to the board and draw their structural formulas.

Sample Questions

- Does the drawing follow the HONC 1234 rule? If not, how can the drawing be altered?
- Can anyone draw this same molecule in a different way?
- Can anyone draw a different molecule with the same molecular formula?
- How does knowing the HONC 1234 rule limit the number of possible structural formulas that you can draw from a particular molecular formula?

Key Point

The HONC 1234 rule is a way to remember the bonding tendencies of hydrogen, oxygen, nitrogen, and carbon atoms in molecules. Hydrogen tends to form one bond, oxygen two, nitrogen three and carbon four.

Discuss Different Ways to Represent the Same Molecular Formula

➡ You might want to pick one molecular formula and stick with it, clarifying the difference between different isomers and different orientations. You could also use 3-D models to clarify.

Sample Questions

- How many isomers does C_3H_9N have? How did you find out?
- How can you tell whether a drawing represents a new molecule?
- Do isomers represent different smells? Explain.
- How can you tell how many structural formulas a molecular formula represents?

Key Point

When trying to decide whether two structures represent the same molecule, you must check how the atoms are connected. Try mentally turning the molecule in space, paying attention to how and where in the molecule different atoms are placed. All these structures represent the same molecule. Only their orientations in space are different.

Wrap-up

Key Question: What are the rules for drawing structural formulas?

- The HONC 1234 rule indicates how many times hydrogen, oxygen, nitrogen, and carbon atoms tend to bond.
- When a molecule is oriented differently in space, it is still the same molecule.

Evaluate (5 minutes)

> **Check-in**
>
> Will any of the molecules shown here have similar smells? Explain your thinking.

Answer: The first and second molecules will smell the same, because they are the same molecule.

Homework

Assign the reading and exercises for Smells Lesson 3 in the student text.

LESSON 4
OVERVIEW

Connect the Dots
Lewis Dot Symbols

Lesson Type
Activity:
Groups of 4

Key Ideas

Nonmetal atoms bond covalently. The bonding tendencies of nonmetal atoms are directly related to the number of valence electrons they have. Lewis dot symbols keep track of the valence electrons of different atoms. In covalent bonds, atoms share pairs of electrons, so only unpaired valence electrons are available for bonding. We can use Lewis dot symbols and the HONC 1234 rule to construct accurate structural formulas.

As a result of this lesson, students will be able to

- create accurate structural formulas using Lewis dot symbols
- describe the type of bonding found in molecular substances
- explain the chemistry behind the HONC 1234 rule

Focus on Understanding

- This latest model—of electrons being paired within atoms—appears to contradict earlier models of the atom, in which electrons are found in a cloud or distributed around a shell. This might confuse students.
- Students might wonder why electrons pair, given that particles with like charges repel each other. Acknowledge that this is strange but is consistent with observations.
- In this lesson students learn "by doing." It is not necessary to thoroughly explain Lewis dot symbols prior to the activity.

Key Terms

Lewis dot symbol
Lewis dot structure
bonded pair
lone pair

What Takes Place

In this lesson, students use Lewis dot symbols and the HONC 1234 rule to create simple molecular structures from molecular formulas. Students are reintroduced to the concept of covalent bonding at the beginning of class. They are provided with Lewis dot manipulative materials to use in creating a variety of molecular structures.

Materials

- student worksheet
- transparencies—ChemCatalyst, Electron Pairs

Per group of 4

- Lewis dot puzzle pieces, only those shown here (7 C atoms, 20 H atoms, 4 O atoms, 1 N atom, 1 S atom, 1 P atom, 1 Cl atom, 4 F atoms, 1 Si atom)

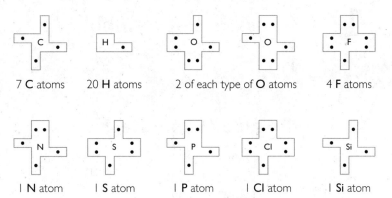

Setup

Place a set of Lewis dot puzzle pieces in a resealable bag for each group of four.

LESSON 4 GUIDE

Connect the Dots
Lewis Dot Symbols

Engage (5 minutes)

Key Question: How does one atom bond to another in a molecule?

ChemCatalyst

These diagrams are called Lewis dot symbols.

·C̈·   ·N̈·   :Ö·   H·

1. What is the relationship between the number of dots, the number of valence electrons, and the HONC 1234 rule?
2. Create a Lewis dot symbol for fluorine, F. How many bonds will fluorine make?

Sample Answers: 1. Carbon has four electrons and makes four bonds. Nitrogen has five electrons and makes three bonds. Oxygen has six electrons and makes two bonds. Hydrogen has one electron and makes one bond. 2. Fluorine has seven valence electrons, so its Lewis dot symbol should show three paired electrons and one unpaired electron. Fluorine will make one bond.

Discuss the ChemCatalyst

➡ Discuss ideas about valence electrons and bonding.

Sample Questions

- How are the Lewis dot symbols related to the number of valence electrons?
- How can you use the Lewis dot symbol to determine the number of bonds an atom will make?
- What symbol did you draw for fluorine, F? How many bonds does fluorine make?
- According to what you've learned so far, what is a covalent bond?

Explore (20 minutes)

Introduce the Activity

➡ Introduce Lewis dot symbols. Draw one or two on the board. Tell students that Lewis dot symbols are a way to keep track of valence electrons. These symbols are extremely useful tools for figuring out bonding and for creating correct molecular structures.

> **Lewis dot symbol:** A diagram that uses dots to show the valence electrons of a single atom.

- Introduce students to the Lewis dot puzzle pieces. Show them that each puzzle piece contains the correct number of valence electrons for that atom. Also show them that each puzzle piece contains the appropriate number of tabs for bonding.

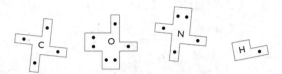

- Tell students that they will be using the puzzle pieces to create molecules according to the worksheet directions. Each group can share a set of puzzle pieces. You can direct students to work in pairs or in groups of four.

LESSON 4 ACTIVITY

Connect the Dots
Lewis Dot Symbols

Name _____
Date _____ Period _____

Purpose
To investigate the role of electrons in covalent bonding.

Materials
- Lewis dot puzzle pieces

Instructions
The puzzle pieces are called Lewis dot symbols. You can use them to pair up electrons to construct models of molecules.

H:N̈:H
Ḧ

Lewis dot structure for NH_3

Part I: Create Molecules

1. Use the puzzle pieces to construct these molecules. Then draw the Lewis dot structure for each molecule, leaving off the outline of each puzzle piece.

   PH_3 HOCl F_2 CH_3Cl

   | H:P̈:H H | H:Ö:C̈l: | :F̈:F̈: | :C̈l: H:C:H H |

2. Use the puzzle pieces to create more molecules here. Draw the Lewis dot structure of each molecule and write the molecular formula below it.

   a. Use one S atom and as many H atoms as you need.
   b. Use one Si atom and as many F atoms as you need.
   c. Use two O atoms and as many H atoms as you need.

   H:S̈:H   :F̈:   H:Ö:Ö:H
              :F̈:Si:F̈:
   H_2S :F̈: H_2O_2
              SiF_4

3. Use the puzzle pieces to construct a molecule with the molecular formula C_2H_6. Draw its Lewis dot structure and its structural formula below.

   H H   H H
   H:C:C:H   H—C—C—H
   H H   H H

126 Unit 2 Smells
Lesson 4 • Worksheet

Living By Chemistry Teaching and Classroom Masters: Units 1–3
© 2010 Key Curriculum Press

4. Use the puzzle pieces to construct all possible isomers of C_3H_8O. Draw Lewis dot structures below. Do the molecules follow the HONC 1234 rule?

```
   H H H              H H   H              H
H:C:C:C:O:H     H:C:C:O:C:H        H:O:H
   H H H              H H   H        H:C:C:C:H
                                          H H H
```

The three isomers do follow the HONC 1234 rule.

5. Use the puzzle pieces to design your own molecule with at least five carbon atoms. Draw its Lewis dot structure. What is the molecular formula of your designer molecule? Does it obey the HONC 1234 rule?

Answers will vary. Note that students might run into trouble with puzzle pieces overlapping each other. This is to be expected sometimes, because molecules are three-dimensional structures.

Part 2: Valence Electrons

Remove one card of each type of atom. Sort these puzzle pieces according to the periodic table.

1. Record your card sort by copying it into the table. Hydrogen and helium have already been done. Include the symbol for the element and the dots.

H·	He:

·C·	·N·	·O·	:F·	:Ne:
·Si·	·P·	·S·	:Cl·	:Ar:

2. List two patterns that you notice in your table.

The Lewis dot symbols in the two rows match. The number of valence electrons goes up from four to eight across each row while the number of bonds goes from four to none and the number of unpaired electrons decreases.

3. **Making Sense** Using what you've learned, explain why the HONC 1234 rule works.

HONC 1234 is an easy way to remember that hydrogen, oxygen, nitrogen, and carbon atoms have one, two, three, and four valence electrons, respectively, that are capable of being shared with other atoms.

4. **If You Finish Early** Draw the Lewis dot structures for two different molecules with the molecular formula C_2H_7N.

```
   H H                    H   H
H:C:C:N:H           H:C:N:C:H
   H H H                H   H H
```

Explain and Elaborate (15 minutes)

Discuss the Use of Lewis Dot Symbols

⇒ Ask students to come to the board to draw the structural formulas of some of the molecules they created.

⇒ Use the HONC 1234 rule to check each molecule. Have students correct them as needed.

Sample Questions

- How do the puzzle pieces keep track of how many bonds each atom makes?
- How many possible isomers does the molecular formula C_3H_8O have? Explain how you figured this out. (It has three.)

Key Point

You can use Lewis dot *symbols* to create Lewis dot *structures*. When you draw an atom using dots to represent the valence electrons, the drawing is called a Lewis dot symbol. A Lewis dot structure is a structural formula of a molecule in which the lines (covalent bonds) have each been replaced with a pair of dots. All the valence electrons on each atom in the molecule are shown as dots. Lewis dot structures make it clear that covalent bonds involve the sharing of a pair of electrons.

> **Lewis dot structure:** A diagram that uses dots to show the valence electrons of a molecule.

Discuss Covalent Bonds and the Pairing Up of Valence Electrons

⇒ Display the transparency Electron Pairs in order to introduce unpaired electrons, lone pairs, and so on.

Sample Questions

- Oxygen has six valence electrons and forms two bonds. Which electrons are involved in the bonding? (the solitary, or unpaired, ones)
- Explain how the number of bonds is related to the numbers of paired and unpaired electrons.
- Explain why the HONC 1234 rule works.

Key Points

A covalent bond is the sharing of a pair of electrons between two nonmetal atoms. This pair of electrons is referred to as a bonded pair. The Lewis dot symbols show us that the number of covalent bonds an atom makes is dependent on how many single, unpaired electrons it has.

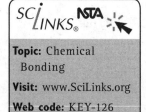

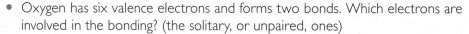

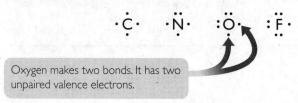

Oxygen makes two bonds. It has two unpaired valence electrons.

Bonded pair: A pair of electrons that are shared in a covalent bond between two atoms.

Some valence electrons are not involved in bonding. The transparency shows the group of atoms that come together to form the structure methanol, CH_4O. The bonded pairs can be seen shared between two atoms. However, notice the pair of electrons not shared between two atoms in the molecule. These are referred to as lone pairs of electrons. They are not involved in bonding within the molecule.

Lone pair: A pair of valence electrons not involved in bonding within a molecule. The two electrons belong to one atom.

Wrap-up

Key Question: How does one atom bond to another in a molecule?

- A covalent bond is a bond in which two atoms share a pair of valence electrons.
- Lewis dot symbols show the valence electrons in an atom and are used to predict bonding in a molecule.
- In a Lewis dot structure, a pair of electrons that are shared in a covalent bond is called a bonded pair. Pairs of electrons that are not involved in bonding and belong to one atom are referred to as lone pairs.
- The HONC 1234 rule indicates how many electrons are available for bonding in atoms of hydrogen, oxygen, nitrogen, and carbon.

Evaluate (5 minutes)

Check-in

The molecular formula $C_4H_{10}O$ has seven different isomers. Draw the structural formula of one of them. You can use your puzzle pieces to assist you.

Answer: Isomers can have a string of four carbon atoms, or CH_3 groups can branch off the main carbon chain, or an oxygen atom between two carbon atoms. Two possible isomers are shown.

Homework

Assign the reading and exercises for Smells Lesson 4 in the student text.

LESSON 5
OVERVIEW

Eight Is Enough
Octet Rule

Lesson Type
Activity:
Groups of 4

Key Ideas

In molecular substances, atoms—with the exception of hydrogen—share electrons in such a way as to obtain a total of eight valence electrons. Thus, the electron arrangement of a covalently bonded atom resembles that of a noble gas. This tendency to achieve eight valence electrons through bonding is referred to as the octet rule. The octet rule also allows for double and triple bonding in covalent compounds.

As a result of this lesson, students will be able to

- apply the octet rule to predict bonding in molecules
- draw Lewis dot structures and structural formulas for molecules that contain double and triple bonds

Key Terms

octet rule   double bond   triple bond

What Takes Place

Students are introduced to the octet rule at the opening of class. While completing a worksheet, students practice creating structural formulas. They problem-solve structures with double and triple bonds.

Materials

- student worksheet

Per group of 4

- Lewis dot puzzle pieces from Lesson 4: Connect the Dots, plus these new pieces.

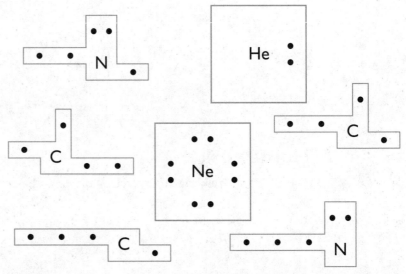

LESSON 5
GUIDE

Eight Is Enough
Octet Rule

Engage (5 minutes)

Key Question: How do atoms bond to form molecules?

> **ChemCatalyst**
>
> Draw the Lewis dot structure for the two covalently bonded molecules shown here. Explain how you arrived at your answer.
>
> **a.** Cl_2 **b.** O_2

Sample Answer: a. The Lewis dot symbol for chlorine has seven dots, so it has only one electron available for bonding. Two chlorine atoms must share a pair of electrons between them. b. Oxygen has six valence electrons, so it has two electrons available for bonding. Perhaps two oxygen atoms share two pairs of electrons.

$$:\ddot{Cl}\cdot \; + \; \cdot\ddot{Cl}: \longrightarrow :\ddot{Cl}:\ddot{Cl}: \qquad :\ddot{O}\cdot \; + \; \cdot\ddot{O}: \longrightarrow :\ddot{O}::\ddot{O}:$$

Discuss the ChemCatalyst

- Guide students toward the octet rule as they discuss their various solutions to the ChemCatalyst.

- Ask a student to draw the Lewis dot structure for Cl_2 on the board. Work with it until it is correct. Circle the eight valence electrons associated with each chlorine atom in Cl_2.

$$:\ddot{Cl}\cdot \; + \; \cdot\ddot{Cl}: \longrightarrow (:\ddot{Cl}(:)\ddot{Cl}:)$$

- Ask students to draw the Lewis dot structure for O_2 on the board. Work with it until it is correct. Circle the eight valence electrons associated with each oxygen atom in O_2.

$$:\ddot{O}\cdot \; + \; \cdot\ddot{O}: \longrightarrow (:\ddot{O}(::)\ddot{O}:)$$

Sample Questions

- How did you arrive at your drawing for Cl_2?
- What other molecules have similar Lewis dot structures? (all the halogens—F_2, Br_2, I_2, At_2)
- How did you arrive at your drawing for O_2?

> **Octet rule:** Nonmetal atoms combine so that each atom has a total of eight valence electrons by sharing electrons.

Section I   Lesson 5 Eight Is Enough

Explore (20 minutes)

Introduce the Activity

- Introduce the octet rule.
- You might want to draw one or two other Lewis dot structures to demonstrate that the octet rule applies to most molecules.

After bonding, each chlorine atom has a total of eight valence electrons surrounding it.

Cl_2 PCl_3 H_2S

Each atom in the molecule has an "octet" of valence electrons. Note that the hydrogen atoms do not follow the octet rule. They have a total of two electrons, similar to the noble gas helium, He.

- Tell the class that they will be using the octet rule to create models of molecules.
- Remind students that each line in a structural formula represents a pair of bonded electrons.
- Pass out the Lewis dot puzzle pieces to each group. Tell students that you have added a few puzzle pieces for making double bonds.

LESSON 5 ACTIVITY

Eight Is Enough
Octet Rule

Name _____
Date _____ Period _____

Purpose
To apply the octet rule to creating Lewis dot structures and structural formulas.

Instructions

1. Fill in the table with the correct structural formulas and Lewis dot structures. Check your drawings against the octet rule and the HONC 1234 rule. Include lone pairs in the structural formulas.

Molecular formula	Structural formula	Lewis dot structure
C_2H_6	H H │ │ H—C—C—H │ │ H H	H H H:C:C:H H H

C_2H_4	H H
	  \\ /
	   C=C
	  / \\
	 H H
		H:C::C:H
		H H

C_2H_2	H—C≡C—H	H:C:::C:H

2. Explain how you can tell from the molecular formula when a compound has a double or a triple bond.

   Molecules made up of carbon and hydrogen with a double bond have fewer H atoms than they would otherwise.

3. Fill in the table with the correct structural formula. Include lone pairs in the structural formulas.

Molecular formula	Structural formula	Molecular formula	Structural formula
H_2	H—H	I_2	:Ï—Ï:
Cl_2	:C̈l—C̈l:	N_2	:N≡N:
O_2	:Ö=Ö:		

4. Molecules that are made up of two atoms are called **diatomic molecules.** Some elements—such as N_2, O_2, and all the halogens—are found as diatomic molecules in nature. Why do you think this is the case?

   These atoms share electrons to achieve an octet. The diatomic molecules resemble noble gases in their electron configuration and are more stable due to this.

5. Fill in the table with the correct molecular formula or structural formula. Include lone pairs in the structural formulas.

Molecular formula	Structural formula	Molecular formula	Structural formula
CH_4O	H–C(H)(H)–Ö–H (with lone pairs)	CH_5N	H–C(H)(H)–N(H)–H (with lone pair)
CH_2O	H–C(H)=Ö (with lone pairs)	CH_3N	H–C(H)=N–H (with lone pair)
CH_4O_2	H–Ö–C(H)(H)–Ö–H (with lone pairs)	HCN	H–C≡N (with lone pair)
CH_2O_2	H–Ö–C(H)=Ö (with lone pairs)	CO_2	Ö=C=Ö (with lone pairs)

6. **Making Sense** Describe the process you use to determine the structure of molecules.

   Answers will vary.

7. **If You Finish Early** From what you've learned so far, predict how the molecules in the table in Question 5 will smell. Explain your predictions.

   Molecules with a nitrogen atom might smell fishy. The molecules with one oxygen atom might smell minty. Molecules with two oxygen atoms might smell sweet. However, carbon dioxide does not have a smell. Also, CH_4O_2, unlike most sweet-smelling compounds, has no double bonds.

Explain and Elaborate (15 minutes)
Discuss the Molecular Structures from the Classwork

→ Ask students to come to the board to draw the various structures from their worksheet.

Sample Questions

- What process did you use in completing the Lewis dot structures?
- How can you use the HONC 1234 rule to check your Lewis dot structures?
- How did you know where to put lone pairs of electrons on the structural formulas?

Key Point

The HONC 1234 rule and the octet rule both help you figure out Lewis dot structures and structural formulas. The HONC 1234 rule allows you to figure out how many bonds there are for each H, O, N, and C atom. For example, nitrogen forms three bonds. Lewis dot structures show that the bonds are pairs of electrons shared between two atoms. In addition, the Lewis dot structure shows lone pairs of electrons. For example, nitrogen has three bonding pairs of electrons and one lone pair.

Discuss Double and Triple Bonding

→ Draw the Lewis dot structure and structural formula of C_2H_2, C_2H_4, and C_2H_6 on the board.

Sample Questions

- When do you need double bonds to satisfy the octet rule? Explain.
- Do molecules with double bonds satisfy the HONC 1234 rule? Explain.
- Is it possible to make a triple-bonded oxygen compound? Explain.
- In theory, are quadruple-bonded carbon compounds possible? Explain.

Key Points

Both the HONC 1234 rule and the octet rule can be satisfied by using double and triple bonds appropriately. Double and triple bonds are sometimes referred to as multiple bonds.

It is not possible to create a triple-bonded oxygen compound, according to the HONC rule. Oxygen has only two unpaired electrons and thus forms only two bonds. A quadruple-bonded carbon compound or silicon compound is possible, but these compounds are not very stable and do not last long.

There are exceptions to the bonding rules laid out here. One notable exception is carbon monoxide, a dangerous gas that has almost no smell. It has a triple bond between the carbon and the oxygen atoms, meaning that it does not satisfy the HONC 1234 rule. However, it does satisfy the octet rule in an unusual fashion. The oxygen atom contributes four electrons to the triple bond, while the carbon atom contributes two.

$$\cdot \overset{\cdot}{\underset{\cdot}{C}} \cdot \text{ and } \cdot \overset{\cdot\cdot}{\underset{\cdot\cdot}{O}} \text{: form } \text{:} C \text{:::} O \text{:}$$

Speculate on the Smells Associated with the Structures in the Classwork

Sample Questions

- What smells did you predict for the various molecules? What was your reasoning?
- Have you smelled any of these compounds? (Carbon dioxide and carbon monoxide are both odorless gases. HCN, hydrogen cyanide, smells like almonds.)

Key Point

Here are a few smells that students might identify. Molecules with a nitrogen atom might smell fishy, molecules with a single oxygen atom might smell minty, and molecules with two oxygen atoms might smell sweet. Carbon monoxide and carbon dioxide do not have a smell. CH_4O_2 has no double bonds, as do most sweet-smelling molecules, so perhaps it does not smell sweet.

Wrap-up

Key Question: How do atoms bond to form molecules?

- Elements form covalent bonds by sharing electrons until each atom has eight valence electrons. This is called the octet rule. Hydrogen is an exception. It forms bonds such that it has two valence electrons.
- Atoms can form double and triple bonds to satisfy the octet rule.
- When covalent bonds form, each atom resembles a noble gas in its electron configuration.

Evaluate (5 minutes)

> **Check-in**
>
> 1. One of these compounds has multiple bonds in it. Which one is it? Explain.
>
> C_4H_{10} C_4H_6
>
> 2. Draw one possible structural formula for C_4H_6.

Answers: 1. C_4H_6. The compound with fewer hydrogen atoms must have multiple bonds. 2. Two possible structural formulas for C_4H_6 are shown (there are others).

$$\text{H}-\underset{|}{\overset{H}{C}}=\underset{H}{\overset{|}{C}}-\underset{|}{\overset{H}{C}}=\underset{H}{\overset{|}{C}}-\text{H} \qquad \text{H}-C\equiv C-\underset{|}{\overset{H}{C}}-\underset{H}{\overset{|}{C}}-\text{H}$$

Homework

Assign the reading and exercises for Smells Lesson 5 in the student text.

LESSON 6 OVERVIEW

Where's the Fun?
Functional Groups

Lesson Type
Activity:
Pairs

Key Ideas

The properties of a molecule are intimately related to common structural features of those molecules, in particular, to functional groups. Smell is one property directly related to functional groups. The presence of functional groups in molecules is also the key to naming molecular compounds.

As a result of this lesson, students will be able to

- identify and name basic functional groups within molecules
- relate certain functional groups to certain smell categories
- describe the naming patterns found among molecules associated with specific functional groups
- deduce the probable smell of a compound from its name or structural formula

Focus on Understanding

- It takes practice to visually distinguish functional groups. For one thing, the differences between some functional groups can be subtle. In addition, a molecule's orientation in space can cause confusion in identifying functional groups.

Key Term

functional group

What Takes Place

This activity introduces functional groups to the class. Pairs of students sort cards containing information on the smell, molecular formula, chemical name, and structural formula of 21 different molecules. They look for patterns in the data that might relate the structure of the molecules of a compound to the compound's smell. After results are shared, the instructor introduces the term *functional group* and clarifies each functional group's specific structure. Finally, a Smell Summary chart is compiled based on the evidence acquired so far.

Note: The Teacher Guide and Student Edition use a combination of IUPAC and common names for molecules. The IUPAC names are more systematic, but in some cases the common name is much shorter or more familiar (e.g., L-carvone rather than its IUPAC name, 2-methyl-5 prop-1-en-2-yl-cyclohex-2-en-l-one.)

SciLINKS NSTA
Topic: Organic Compounds
Visit: www.SciLinks.org
Web code: KEY-206

Materials

- student worksheet
- transparencies—ChemCatalyst, Ketones, Amines, Carboxylic Acids, Esters, Alkanes

- large sheet of paper (approximately 3 ft by 3 ft) and markers for Smell Summary chart

Per pair

- Structural Formula cards—one set, 21 cards in each set

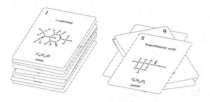

Setup

Prepare card sets to pass out to pairs of students. There should be 21 cards in each set.

LESSON 6 GUIDE

Where's the Fun?
Functional Groups

Engage (5 minutes)

Key Question: What does the structure of a molecule have to do with smell?

ChemCatalyst

Consider the following compounds. List at least three differences and three similarities between the two molecules.

```
    H  O         H  H              H  H  H  O
    |  ||        |  |              |  |  |  ||
H—C—C—O—C—C—H         H—C—C—C—C—O—H
    |            |  |              |  |  |
    H            H  H              H  H  H
       Molecule 1                  Molecule 2
```

Sample Answer: Differences: The structures are different. The double-bonded oxygen atom is connected to the second carbon atom in one molecule and to the fourth carbon atom in the other. The oxygen atoms are in the middle of one molecule and near the end of the other molecule. **Similarities:** They are composed of the same elements. Both have the same molecular formula, $C_4H_8O_2$. Both have two oxygen atoms and a double-bonded oxygen atom. They are similar in shape.

Discuss the ChemCatalyst

➡ Begin to draw the connection between molecular structure and smell.

Sample Questions

- What are the major differences in the structures of these molecules?
- How do you think these two molecules smell? Would you be surprised to learn that they have different smells? Why or why not?
- Do you think there is a relationship between the structure of a molecule and its properties? Explain your thinking.

Explore (15 minutes)

Introduce the Activity

➡ Hand out Structural Formula cards to each pair to sort according to the directions on the worksheet.

1

L-carvone

[structural formula of L-carvone]

$C_{10}H_{14}O$

minty

Section I Lesson 6 Where's the Fun? 41

LESSON 6 ACTIVITY

Where's the Fun?
Functional Groups

Name _____
Date _____ Period _____

Purpose
To explore the connection between molecular structure and smell.

Materials
- 1 set of 21 structural formula cards

Questions

1. Find all the molecules that have two oxygen atoms. What patterns do you discover among these molecules?

   The molecules that contain two oxygen atoms are either sweet smelling or putrid smelling. They look similar, but the oxygen atoms are always on the end of the putrid-smelling molecules and in the middle of the sweet-smelling molecules.

2. Find all the molecules that have a ring structure. Do all these molecules smell the same? Explain what you discovered.

   Most of the molecules with ring structures smell minty. However, two of them don't. One is a fishy molecule and one is putrid smelling.

3. Sort the molecules based on the number of carbon atoms they have. Do you think the smell of a molecule is related to the number of carbon atoms it has? Explain.

   The sweet smelling molecules have different numbers of carbon atoms. So the number of carbon atoms doesn't seem to predict smell.

4. Sort all the molecules according to their names. What patterns do you discover?

   Sorting by name also sorts these molecules by smell and structural features (such as the presence of a nitrogen atom in the fishy-smelling molecules, rings in the minty-smelling molecules, and a double-bonded oxygen atom in the minty-smelling molecules).

5. Sort all the molecules by smell. What structural features do the molecules in the groups below have in common? Draw (or describe) the features.

Fishy-smelling molecules

nitrogen atom

$$-\overset{|}{\underset{|}{C}}-\overset{|}{\underset{|}{N}}-$$

Sweet-smelling molecules

two oxygen atoms bonded a certain way within the molecules

$$-\overset{O}{\overset{\|}{C}}-O-\overset{|}{\underset{|}{C}}-$$

Putrid-smelling molecules

two oxygen atoms bonded a certain way on the end of the molecules; an —OH group on the end

$$-\overset{O}{\overset{\|}{C}}-O-H$$

Gasoline-smelling molecules

Made up of only carbon and hydrogen atoms (Note: Small alkanes don't smell. This will be covered in Lesson 19: Sniffing It Out.)

$$-\overset{|}{\underset{|}{C}}-$$

Minty-smelling molecules

A single oxygen atom double bonded to a carbon atom; a ring structure.

$$-\overset{|}{\underset{|}{C}}-\overset{O}{\overset{\|}{C}}-\overset{|}{\underset{|}{C}}-$$

6. Can you think of another way to sort the molecule cards? Do any of these other sorting methods help you predict smell?

Answers may vary. You could sort by the number of double bonds, or by the number of hydrogen atoms. However, these other sorting methods do not help predict smell.

7. What smell would you predict for this molecule? Provide evidence to support your prediction.

$$H-\overset{}{\underset{\overset{\|}{O}}{C}}-O-\overset{H}{\underset{H}{C}}-H$$

Based on patterns seen in the structures of the other molecules, this molecule should smell sweet. It contains a structure similar to that of all the other sweet-smelling molecules.

8. **Making Sense** What structural features seem to be the best predictors of the smell of a molecule? Be specific.

The presence of atoms other than carbon and hydrogen and the presence of specific structural features seem to be the best predictors so far.

Explain and Elaborate (20 minutes)
Discuss the Patterns Found in the Molecular Structures

➡ Make a list of students' generalizations.
➡ Allow students to assist you in identifying the common structural features in molecules that smell the same. Have students draw these features on the board.

Sample Questions

- What were some of the discoveries you made as a result of sorting the cards?
- Do your best to describe the structural features that all the sweet-smelling molecules have in common.
- What structural features do you find in all the fishy-smelling molecules?
- What do the molecules ending in "-ane" have in common?
- What do all the molecules that smell like gasoline have in common?

Key Points

A number of patterns emerge from the card sort. For example, the smell and the name of a molecule are directly connected. Also, both the sweet-smelling molecules and the putrid-smelling molecules have two oxygen atoms. The minty-smelling molecules all contain a ring and have a double-bonded oxygen atom. If the cards are sorted by smell, you will find five groups: fishy, minty, sweet, putrid, and gasoline smell. Sorting by name should produce the same outcome.

Each group of molecules with a similar smell has something identical in its structure. All the sweet-smelling molecules, for example, contain an identical piece in them: an oxygen atom connected to two carbon atoms, one of which is bonded to a second oxygen atom. The other parts of the sweet-smelling molecules may be quite different from each other, but every sweet-smelling molecule in our sample contains this same feature.

$$\begin{array}{c} \text{O} \\ \| \\ -\text{C}-\text{O}-\text{C}- \end{array} \qquad \begin{array}{c} \text{O} \\ \| \\ -\text{C}-\text{C}-\text{C}- \end{array}$$

This feature is found in all the sweet-smelling molecules in our sample.

This feature is found in all the minty-smelling molecules in our sample.

You can see that minty-smelling molecules all have an identical structural feature as well: a carbon atom connected to a single oxygen atom by a double bond.

These specific features are referred to as functional groups.

> **Functional group:** A cluster of atoms in a molecule that is responsible for many of its properties.

Introduce Functional Groups

➡ Display the transparencies showing the molecules sorted according to smell.
➡ You might draw and label each functional group separately on the board for clarity.

Sample Questions

- What functional group is found in all the fishy-smelling molecules? The putrid-smelling molecules?

- How are the putrid-smelling molecules and the sweet-smelling molecules similar? How are they different?
- What generalization can you make about the molecules that smell like gasoline?
- How are the names of the molecules related to their functional groups?

Key Points

The functional groups have names, and molecules frequently are named according to the functional groups they contain.

Ketone group

Carboxyl group

Ester group

Amine group

Functional groups can be shown without the rest of the molecule.

Ketone group

Carboxyl group

Ester group

Amine group

All the molecules containing ester functional groups end in "-ate." All the ketones end in "-one." All the amines end in "-ine."

There are other functional groups not covered here.

Alcohols are molecules that all contain a hydroxyl group (–OH). The names of alcohols all end in "-ol," for example, ethanol or propanol. Students will discover more about the smells of alcohols in later lessons.

Hydroxyl group

Molecules that smell like gasoline don't seem to have any particular functional group. These molecules are called alkanes, and they are made entirely of carbon and hydrogen atoms.

Alkane

The structure of an aldehyde is similar to that of a ketone, except that the carbon atom that is double bonded to the oxygen atom is attached directly to at least one hydrogen atom. In the structure of "formaldehyde," the carbon atom with the double bonded oxygen is attached directly to two hydrogen atoms.

$$\begin{array}{c} O \\ \parallel \\ -C-H \end{array}$$

Aldehyde group

An ether has a single oxygen atom located between two carbon atoms.

$$-C-O-C-$$

Ether group

There are exceptions to the patterns related to smell. For example, small alkanes such as methane, CH_4, or propane, C_3H_8, have no smell.

Create a Smells Summary Chart

➡ Tape up a large piece of paper (about 3 ft by 3 ft). Create a poster that summarizes the class's thinking so far.

➡ Solicit information about how molecular formula, chemical name, and functional group are related to smell.

Sample Questions

- Is the molecular formula of a molecule ever helpful in predicting smell? If so, when?
- What specific information in the chemical name helps you predict smell?
- How can you use a functional group to predict smell?
- Do any other features or properties help you predict smell?
- Do we need to go further in our investigation in order to understand the chemistry of smell? Explain your thinking.

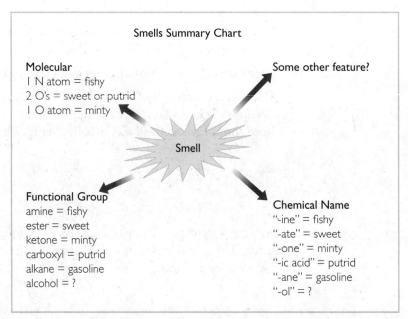

Wrap-up

Key Question: What does the structure of a molecule have to do with smell?

- Functional groups are structural features that show up repeatedly in molecules and seem to account for some of their chemical properties.
- Molecules containing the same functional groups have similar smells.
- The chemical names of molecules are often related to the functional groups they contain. Thus, chemical name is related to smell.

Evaluate (5 minutes)

> **Check-in**
>
> If a molecule is sweet-smelling, what other things do you know about it? List at least three things that are probably true.

Answer: It probably contains two oxygen atoms and an ester functional group. One of the oxygen atoms is double bonded. The chemical name probably ends in "-ate."

Homework

Assign the reading and exercises for Smells Lesson 6 in the student text. Tell students to be sure to do the Lab Prep exercise, which is a prelab for the next class. Tell them to be prepared to pass a lab-readiness quiz next time before they begin the lab.

LESSON 7 OVERVIEW

Create a Smell
Ester Synthesis

Key Ideas

Lesson Type
Lab:
  Pairs

It is possible for chemists to create different smells in the laboratory by creating molecules with specific functional groups. For instance, an alcohol and a carboxylic acid can be induced to react to create a new molecule that smells sweet. The different smells can be attributed to specific sequences of atoms, or functional groups.

Note: The terms *chemical reaction* and *synthesis* are covered more thoroughly in Lesson 7 of the Student Edition.

As a result of this lesson, students will be able to

- successfully complete a laboratory procedure to produce sweet-smelling esters

Key Term

synthesis

What Takes Place

This class is a formal lab experiment that results in the formation of three esters (sweet-smelling) from various organic acids and alcohols. The ChemCatalyst is designed to check for lab-readiness. Before the lab, address the class as a whole on the basic steps of the experiment and on safety concerns. A worksheet provides lab instructions and questions to answer. Students work in pairs to set up their equipment and complete the procedure. The lab ends with clean-up and a sharing of preliminary results.

This lab is processed more thoroughly in Lesson 8: Making Scents.

Materials

- student worksheet
- safety goggles
- gloves and aprons (recommended)

Per pair

- 50 mL beaker
- hot plate
- 3 microscale test tubes (0.5 mL)
- microscale test tube rack (or 1 more 50 mL beaker)
- 3 plastic pipettes
- scissors to cut pipettes

- boiling stones
- waste container for disposal of esters

Per group of 4
- water, 60 mL
- isopentanol, $C_5H_{12}O$, 2 mL
- butanol, $C_4H_{10}O$, 2 mL
- ethanol, C_2H_6O, 2 mL
- glacial acetic acid, $C_2H_4O_2$, 2 mL
- butyric acid, $C_4H_8O_2$, 1 mL
- 18 M sulfuric acid, H_2SO_4, 1 mL
- 6 dropper pipettes (for acids and alcohols)
- baking soda for spills
- pencil or marker for labeling test tubes

Setup

Arrange the chemicals in reagent bottles at stations. Set up waste containers for the esters at each station. Have baking soda available to neutralize acid spills. You may choose to distribute the more concentrated acids yourself rather than have the students handle them.

> ### Safety Precautions
>
> There should be **no open flames** in this lab, because the organic chemicals are flammable.
>
> All bottles or containers should be closed when not in use to lessen fire danger and to minimize odors in the room.
>
> Extreme care should be taken when handling 18 M sulfuric acid. Have baking soda available to neutralize spills.
>
> Heating must be done slowly and carefully.
>
> Butyric acid is extremely pungent. Take extra precautions to avoid spills and drips.

Cleanup

Put boiling stones and used pipettes in the trash. Have students pour ester products into a common waste receptacle. Place leftover esters in a safe place, allow the liquid to evaporate, and dispose of the solids at an appropriate waste disposal site. Stopper and save remaining alcohol and acid reagents.

LESSON 7 GUIDE

Create a Smell
Ester Synthesis

Engage (5 minutes)

Key Question: How can a molecule be changed into a different molecule by using chemistry?

> **ChemCatalyst**
> 1. What are some of the starting ingredients you will be using in this lab?
> 2. Name something you will be doing to the chemicals in this experiment.

Answers will vary.

Check Students for Lab Readiness

➡ If answers are incorrect, have students reread the lab until they can provide correct answers.

Explore (25 minutes)

Introduce the Lab

➡ Go over safety guidelines with the class.
- Everyone must wear safety goggles.
- You will be using a hot plate (medium heat) to heat the ingredients. There should be **no open flames,** because several of the chemicals are flammable.
- Remember to waft the chemicals when you want to smell them. Some of the chemicals you are using smell very bad.
- All uncovered bottles of alcohols and esters should be kept far away from flames, because they are extremely flammable.
- Concentrated sulfuric acid will also be used in the reactions. It is very caustic and can burn the skin. Baking soda is available to neutralize any spills.

➡ Pass out the worksheet. Ask students to work in pairs. Tell students where they can pick up the equipment and chemicals in your particular classroom.

➡ Emphasize the importance of exact measurements. If the reactants are not mixed in the appropriate proportions, leftover carboxylic acid might mask any ester that has been created, producing a stinky outcome instead of a sweet one.

➡ Show students an example of the apparatus they will be using. Go over the important points of the lab as outlined here.
- The water should be at a gentle boil.
- The three test tubes must be labeled 1, 2, and 3.
- Make certain you carefully waft the starting chemicals to smell them.
- Smell the mixtures again before heating.
- Record any errors that crop up during the experiment (e.g., putting too many drops of sulfuric acid into the tube). These may be important in processing the lab later.
- If the mixture still smells putrid after heating, resume heating for another 5 minutes or until the smell changes.

LESSON 7 LAB

Create a Smell
Ester Synthesis

Name _____
Date _____ Period _____

Purpose
To create new smells by following a formal laboratory procedure.

Materials
- 50 mL beaker
- hot plate
- 3 microscale test tubes
- boiling stones
- 3 plastic pipettes
- scissors to cut pipettes
- pencil or marker
- organic acids and alcohols (see below)

Setup and Safety Instructions

 Always wear safety goggles and dress appropriately for a chemistry lab.

There should be NO OPEN FLAMES. The organic chemicals are flammable.

Extreme care should be taken when handling 18 M sulfuric acid. It burns the skin and creates holes in clothing.

During clean-up, place the final products in designated waste containers.

Recap bottles after use to reduce unwanted odors in the room.

Procedure

1. Fill a 50 mL beaker with about 30 mL of water.
2. Drop in a boiling stone.
3. Place the beaker of water on a hot plate and bring the water to a *gentle* boil.
4. Label your test tubes 1, 2, and 3.
5. Carefully smell the acids by wafting, and record the smells in the data table. Add **five** drops of the appropriate carboxylic acid to each test tube according to the table.

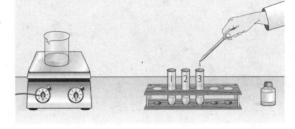

6. Carefully smell the alcohols by wafting, and record the smells in the data table. Add **ten** drops of the appropriate alcohol to each tube according to the table.

Test tube	Carboxylic acid	Alcohol
1	acetic acid	isopentanol
2	acetic acid	butanol
3	butyric acid	ethanol

7. Add **one** drop of the concentrated sulfuric acid, H_2SO_4, to each test tube.
8. Drop a boiling stone into the mixture in each tube.
9. Carefully smell each mixture by wafting, and record the smells in the data table.

Data Table

Test tube	Smell of carboxylic acid	Smell of alcohol	Smell of mixture before heating	Smell of mixture after heating
1	vinegary	medicinal	putrid	sweet, banana
2	vinegary	medicinal	putrid	sweet, fruity
3	like rotten cheese	medicinal	putrid	sweet, pineapple

10. Cut a plastic pipette so that it is shorter than the length of the test tube. Put it in the test tube with the stem down, so that the bulb *loosely* seals off the tube.

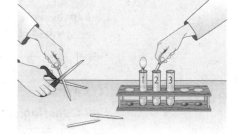

11. Place the test tubes into the boiling water and heat for five minutes, or until the smell is no longer putrid.
12. After five minutes, remove the test tubes from the water. Turn off hot plate.
13. Remove the pipette from the test tube. It should not have any liquid in it, just vapors.
14. Carefully squeeze the pipette near your nose so that you can waft and smell the vapors. Record the smells in the data table.
15. Dispose of chemicals as directed by your teacher. Turn off hot plate.
16. **Making Sense** What functional group do you think is present in the final molecules? Explain.

    The sweet smells of the final products suggest the presence of an ester functional group in these molecules.

17. What do you think happened to the molecules to change the smell?

    Answers will vary. The original molecules were changed into different molecules with different properties. Perhaps the acid and the alcohol joined together somehow.

Explain and Elaborate (10 minutes)

Gather the Class Data

➡ Draw the data table on the board and collect the results of the experiment as a class.

➡ Make sure students have all the information they need to complete a formal lab write-up as homework.

Sample Questions

- How did the smells after heating compare to those before heating? Did you notice any patterns?
- What functional group do you think is present in the final molecules? Explain.
- What happened to the molecules to change the smell?

Key Point

In this lab you used chemistry to produce new molecules with new properties. A chemical change took place as the result of a chemical reaction. When chemists purposely produce a specific compound, the process is called chemical synthesis.

> **Synthesis:** The creation of specific compounds by chemists through controlled chemical reactions.

Wrap-up

Key Question: How can a molecule be changed into a different molecule by using chemistry?

- A chemical reaction is the process that results in chemical change, producing new compounds with new properties.
- Two molecules with their own unique smells can combine to produce a new molecule with a different smell.

Evaluate (5 minutes)

There is no Check-in for this class. The lab will be processed in detail in the next lesson.

Homework

Assign the reading and exercises for Smells Lesson 7 in the student text.

LESSON 8 OVERVIEW

Making Scents
Analyzing Ester Synthesis

Lesson Type
Follow-up:
Individuals

Key Ideas

In the ester lab, a chemical change, or chemical reaction, took place. The starting ingredients of a chemical reaction are called the reactants. The ending substances are called the products. During chemical reactions, new compounds with new properties are formed. On a molecular level, a chemical change is a rearrangement of atoms that involves the breaking and making of chemical bonds.

As a result of this lesson, students will be able to

- explain what happened at a molecular level during the ester synthesis lab
- predict the product of a reaction between an alcohol and a carboxylic acid
- generally define a chemical reaction
- define what a catalyst is

Key Terms

chemical equation
reactant
product
catalyst

What Takes Place

This lesson processes the previous lab in detail and begins to develop the concept of a chemical reaction. The worksheet provides students with the structural formulas of the reactants. Students are guided to come up with the molecular structures of the ester products they produced in the previous class. Students gain a conceptual understanding of what happened to the specific molecules they mixed and heated in the lab. They are introduced to a rudimentary chemical equation.

The formal chemical equations have been purposely left out so students can focus on how the molecular *structures* change during the reaction.

Chemical reactions are handled formally in the Toxins unit after students have had experience with moles in the Weather unit. The purpose of this lesson is to provide a conceptual foundation for understanding chemical reactions.

Materials

- student worksheet
- transparency—Data Table

LESSON 8 GUIDE

Making Scents
Analyzing Ester Synthesis

Engage (5 minutes)

Key Question: What happened to the molecules during the creation of a new smell?

> **ChemCatalyst**
>
> What do you think happened in the experiment in the previous class to transform an acid molecule and an alcohol molecule into a sweet-smelling molecule?

Sample Answer: Students might say that the molecules in the test tube exchanged pieces or that the molecules came apart entirely and rearranged.

Discuss the ChemCatalyst

Sample Questions

- What do you think took place in the test tube when you mixed and heated the initial substances?
- What evidence do you have to support your answer?

Explore (15 minutes)

Introduce the Follow-up

➡ Introduce students to a rudimentary chemical equation. It is not necessary to use chemical formulas yet.

- Chemists show what happens during a chemical reaction with a chemical sentence written as a chemical equation. It might look like this, where A and B are the starting ingredients and the new substances produced are C and D, or just E.

$$A + B \longrightarrow C + D \quad \text{or} \quad A + B \longrightarrow E$$

> **Chemical equation:** A chemical sentence that tracks what happens during a change in matter. Chemical equations are written with chemical formulas and keep track of the atoms involved in the changes.

➡ Use the Data Table transparency to display the results of the lab experiment. This will serve as a reference as students complete the worksheet.

Data Table

Test tube	Organic acid	Alcohol	Smell of mixture before heating	Smell of mixture after heating
1	acetic acid	isopentanol	putrid	fruity, banana smell
2	acetic acid	butanol	strongly putrid	fruity, pear smell
3	butyric acid	ethanol	putrid	fruity, pineapple smell

➡ Have students work individually.

LESSON 8 FOLLOW-UP

Making Scents
Analyzing Ester Synthesis

Name _____
Date _____ Period _____

Purpose
To analyze the results of the ester synthesis lab.

Analysis

1. In the ester synthesis lab, how did the smell of the mixtures before heating compare to the smell of the mixtures after heating?

   They were putrid before heating and sweet smelling after heating.

2. Based on the smell of the mixtures after heating, what functional group must be present in the final molecules that were produced? Draw it.

   An ester functional group must be present.

   $$-\underset{}{\overset{O}{\underset{\|}{C}}}-O-C-$$

3. Using these two structural formulas, build a new molecule that contains the functional group you identified in Question 2. This is a chemical reaction, so you are allowed to break bonds and make new ones.

   Acetic acid   Butanol   **Butyl acetate**

4. Are there any atoms that were not used to make the sweet-smelling molecule in Question 3? If so, what molecule do these pieces make?

   Two hydrogen atoms and an oxygen atom were not used. They make water, H_2O.

5. Complete this chemical equation. Make sure that the equation is balanced (the same number of carbon, hydrogen, and oxygen atoms on both sides of the arrow).

   $C_2H_4O_2 + C_4H_{10}O \xrightarrow{H_2SO_4} H_2O + C_6H_{12}O_2$

6. What evidence do you have that this reaction took place in your test tube?

   An ester must have been produced, because the product smelled sweet. Also, the right number and type of atoms are present in the product.

7. The reaction between acetic acid and isopentanol produces a sweet smell. Draw the structural formulas of the products, water and isopentyl acetate.

[Structural formulas: Acetic acid + Isopentanol → H–O–H + Isopentyl acetate]

8. The reaction between butyric acid ($C_4H_8O_2$) and ethanol (C_2H_6O) produces a sweet smell. Draw the structural formulas of the products, water and ethyl butyrate.

[Structural formulas: Butyric acid + Ethanol → H–O–H + Ethyl butyrate]

9. Imagine that you used the following acid and alcohol in the lab to create a sweet-smelling molecule. Draw the structural formulas of the products, water and octyl formate.

[Structural formulas: Formic acid + Octanol → H–O–H + Octyl formate]

10. What are the molecular formulas of the sweet-smelling products in Questions 5 and 8? Draw the structural formulas of these two molecules next to each other. Why do you think the molecules in Questions 5 and 8 smell different?

[Structural formulas: Butyl acetate and Ethyl butyrate]

Both molecules have the molecular formula $C_6H_{12}O_2$. Differences in smell may be due to the different locations of the ester functional group in the molecules.

11. **Making Sense** Use your own words to describe what happens on a *molecular* level when an acid and an alcohol react.

    Answers will vary.

12. **If You Finish Early** See if you can figure out how the products of these reactions are named. What would be the name of the product in Question 3?

    The names of the reactants are combined, with the alcohol first; then "-ate" is added to the acid. The ester products in Question 3 is butyl acetate and octyl formate.

Explain and Elaborate (20 minutes)

Process the Structures of the Ester Products

➠ Ask students to draw the structures from Exercises 3, 5, 7, and 8 on the board. If a structure is incorrect, guide the students to fix it using the HONC 1234 rule or conservation of matter.

Sample Questions

- How did you decide what the product molecules should look like?
- Are the same atoms present before and after the chemical reaction?

Key Points

The products of these reactions smell sweet, so they must all contain an ester functional group. The atoms are not destroyed and must all be accounted for in the end. The final structure is arrived at by combining the two original molecules and making sure an ester functional group is present. Three atoms are left over, two H atoms and an O atom, which form water.

Many different acids and alcohols can be brought together to form an ester and water. The general description of this reaction is

$$\text{acid} + \text{alcohol} \longrightarrow \text{water} + \text{ester}$$

Assist Students in Making Sense of the Chemical Reactions

➠ Draw one of the chemical reactions in its entirety on the board. During your discussion, draw a box around the atoms that break off to form water.

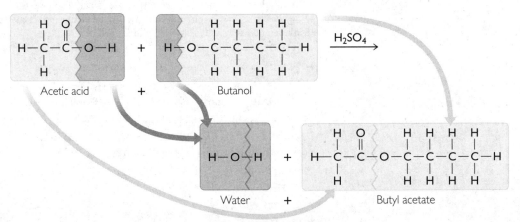

Sample Questions

- What are the products in this equation?
- What part of each molecule changed during the chemical reaction? Which atoms created the water molecule?
- Why do you think H_2SO_4 is shown above the arrow?
- According to our drawing, how many bonds were broken in this reaction? What new bonds were formed?

Key Points

The lab procedure you completed resulted in a chemical reaction. When chemical reactions take place, new compounds with new properties are produced.

During chemical reactions, bonds are broken and the atoms rearrange themselves to form new compounds.

It is possible to track the changes to the structure of the molecules through chemical equations. In a chemical equation, the substances to the left of the arrow are the substances that are mixed and are called the reactants. The new substances that are produced are shown to the right side of the arrow and are called the products. Sometimes a catalyst is added to assist a chemical reaction. This substance speeds up or facilitates the reaction without being in any way changed. A catalyst is not considered a reactant.

> **Reactant:** An element or compound that is a starting ingredient in a chemical reaction. Reactants are written to the left of the arrow in a chemical equation.
>
> **Product:** An element or compound that results from a chemical reaction. Products are written to the right of the arrow in a chemical equation.
>
> **Catalyst:** A substance that accelerates a chemical reaction but is itself not permanently consumed or altered by the reaction. A catalyst is written above the arrow in a chemical equation.

When atoms are rearranged during chemical reactions, not all of the bonds must break. Notice that most of the molecule stays together. In this reaction, only one bond breaks in each molecule. Two new bonds form. Usually, chemical reactions happen in the area of a functional group on a molecule. A bond may break within or next to a functional group, and new bonds then form with other pieces of molecules.

Discuss the Patterns Found in the Chemical Names (optional)

➡ Write out several of the chemical reactions as sentences. Show students the naming patterns.

Sample Questions

- What do the chemical names of the reactants in Question 5 have in common with the chemical names of the products?
- What patterns do you see in the naming of the ester products?

Key Point

The naming of chemical compounds is not random. There is a pattern to it. When butyric acid reacts with ethanol, it forms ethyl butyrate. When formic acid reacts with octanol, it forms octyl formate. The first part of the name of the acid becomes the second part of the name of the ester that is formed. The first part of the name of the alcohol becomes the first part of the name of the ester.

Formic acid reacts with octanol to form octyl formate.

From this pattern, we can correctly predict that the product of acetic acid and propanol will be propyl acetate.

Wrap-up

Key Question: What happened to the molecules during the creation of a new smell?

- The smell of the molecules in the ester lab changed because the reactant molecules combined to form different product molecules.
- In a chemical reaction, bonds are broken and new bonds are formed.
- A catalyst is a substance that accelerates a chemical reaction but is itself not permanently consumed or altered by the reaction.

Evaluate (5 minutes)

Check-in

1. Predict the structural formula of the product of this reaction.

$$\underset{\text{Formic acid}}{\text{H}-\overset{\overset{\text{O}}{\|}}{\text{C}}-\text{O}-\text{H}} \quad + \quad \underset{\text{Ethanol}}{\text{H}-\text{O}-\overset{\overset{\text{H}}{|}}{\underset{\underset{\text{H}}{|}}{\text{C}}}-\overset{\overset{\text{H}}{|}}{\underset{\underset{\text{H}}{|}}{\text{C}}}-\text{H}} \quad \longrightarrow$$

2. What smell would you expect the product to have?

Answers: 1. The structure of this molecule is shown below. Water is also a product.

$$\underset{\text{Ethyl formate}}{\text{H}-\overset{\overset{\text{O}}{\|}}{\text{C}}-\text{O}-\overset{\overset{\text{H}}{|}}{\underset{\underset{\text{H}}{|}}{\text{C}}}-\overset{\overset{\text{H}}{|}}{\underset{\underset{\text{H}}{|}}{\text{C}}}-\text{H}}$$

2. Because ethyl formate is an ester, you would expect it to smell sweet.

Homework

Assign the reading and exercises for Smells Lesson 8 in the student text. Students complete their Ester Synthesis lab report in Exercise 1. Assign the Section I Summary.

SECTION II

Building Molecules

Section II explores molecular shape and the role of electron pairs. Lesson 9 introduces three-dimensional ball-and-stick models. Some new smells force students to rethink their previous hypothesis linking smell to functional groups. In order to help students understand overall molecular shape, they are introduced to electron domain theory and the geometries of small molecules in Lessons 10 and 11. These lessons highlight the role of electrons—both bonded pairs and lone pairs—in determining the shapes of molecules. After using what they've learned to construct some molecules, students move to space-filling models in Lesson 12. In Lesson 13 they come up with generalizations that link molecular shape and smell. Finally, in Lesson 14 they integrate their learning into a model of how the nose works, and the receptor site model is introduced.

In this section, students will learn

- how to interpret three-dimensional models of molecules
- how electron pairs determine molecular geometry
- how to use Lewis dot structures to predict possible molecular structures from molecular formulas
- general relationships between molecular shape and smell
- the receptor site or "lock and key" theory of molecular interaction

LESSON 9 OVERVIEW

New Smells, New Ideas
Ball-and-Stick Models

Lesson Type
Activity:
Groups of 4

Key Ideas

Structural formula and functional group are related to the properties of a molecule, such as smell. However, the overall shape of a molecule can also account for differences in smell. Chemists use ball-and-stick models as three-dimensional representations of molecules to allow them to explore the overall shape of a molecule.

As a result of this lesson, students will be able to

- visually interpret three-dimensional ball-and-stick molecular representations
- translate between molecular models, molecular formulas, and structural formulas
- describe connections between molecular properties and molecular structure

Focus on Understanding

- Students gain valuable conceptual understanding from firsthand contact with the molecular models. The more models available the better.
- Students may or may not notice the specific structural features we've mentioned here, or they may describe them differently. It's okay to allow them their current theories.

Key Term

ball-and-stick model

What Takes Place

Five new smells are sampled in this class. These compounds are all alcohols, but they belong in three different smell categories (sweet, camphor, and minty). Students examine the molecular and structural formulas of these molecules and attempt to explain how molecules with the same functional group might belong to completely different smell categories. They discover that the overall shape of a molecular compound also affects the way it smells.

Materials (sufficient for 1 or more classes)

- transparency—ChemCatalyst
- student worksheet
- 5 plastic pipettes
- rose perfume oil, 5 mL
- ball-and-stick models of citronellol, menthol, and fenchol
- pine cleaner, 5 mL
- jasmine perfume oil, 5 mL
- mint flavor extract, 5 mL
- pine oil, 5 mL

64 *Living By Chemistry Teacher Guide*   Unit 2 Smells

Per group of 4
- 5 cotton balls
- vials I–M

Setup

Prepare sets of vials I–M by first placing a cotton ball in each vial and then using the plastic pipettes to deliver three to five drops of the stock smell solutions to the appropriately lettered vials. Use a new pipette for each essence. You should prepare the vials in a hood or outdoors. Place each set of the five vials in a plastic sandwich bag to make it easy to distribute them to groups of students.

Label	Contents	Smell
I	rose perfume	sweet
J	pine cleaner	camphor
K	jasmine perfume oil	sweet
L	mint flower extract	minty
M	pine oil	camphor

Build ball-and-stick models of citronellol, menthol, and fenchol. Label the models 1, 2, and 3. Do not label the molecules with their chemical names or smells. Build one set for the class to view; or, if you have enough modeling materials, build a set for each group, or enough for groups to share.

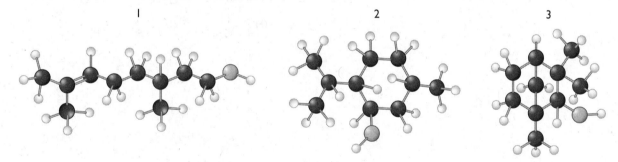

Cleanup

Save the vials for reuse in other classes and over the course of this unit. Within a few weeks, however, you will need to remove the cotton balls and air out the vials; otherwise, the substances' smells will mix with one another and begin to smell putrid.

When you are finished using vials I–M in all your classes, remove the cotton balls from the vials, place them in a plastic bag, and dispose of them. Let the vials air out in a hood or rinse them with acetone for reuse the next time you do this unit.

LESSON 9 GUIDE

New Smells, New Ideas
Ball-and-Stick Models

Engage (5 minutes)

Key Question: What three-dimensional features of a molecule are important in predicting smell?

ChemCatalyst

Do you think any of these molecules will smell similar? What evidence do you have to support your prediction?

citronellol $C_{10}H_{20}O$

geraniol $C_{10}H_{18}O$

menthol $C_{10}H_{20}O$

Sample Answer: Students might predict that these molecules will all smell minty because they have one oxygen atom. Or that the ones that look similar have similar smells. *Note:* Do not give away the smell categories in advance. Students will discover these in today's activity.

Discuss the ChemCatalyst

➦ Discuss the possible smells of the three compounds.

Sample Questions

- What do the molecules have in common?
- Do you think any of these compounds will smell similar? Give your reasoning.
- Suppose you are told that two of the compounds smell sweet. Which two would you predict are the sweet-smelling ones? Explain your reasoning.

Explore (20 minutes)
Introduce the Activity

- Pass out vials I–M to groups of four students. Remind students of the correct procedure for smelling, and tell them to be sure to replace the caps and tighten them when they finish. After sampling is complete, collect the vials.
- Pass out worksheets. When students have completed Part 1 individually, pass out the ball-and-stick models to groups of four for Part 2.

LESSON 9 ACTIVITY

New Smells, New Ideas
Ball-and-Stick Models

Name _____
Date _____ Period _____

Purpose
To examine ball-and-stick models of molecules and compare them to structural formulas.

Part I: New Smell Molecules
Test the samples for smell. Fill in the table, then answer the questions.

Vial	Molecular formula and name	Functional group	Compound type	Structural formula	Smell
I	$C_{10}H_{20}O$ citronellol	hydroxyl group	alcohol		sweet, floral, lemon
J	$C_{10}H_{18}O$ fenchol	hydroxyl group	alcohol		camphor, pine

Living By Chemistry Teaching and Classroom Masters: Units 1–3
© 2010 Key Curriculum Press

Unit 2 Smells 145
Lesson 9 • Worksheet

Vial	Molecular formula and name	Functional group	Compound type	Structural formula	Smell
K	$C_{10}H_{18}O$ geraniol	hydroxyl group	alcohol		sweet, floral, roses
L	$C_{10}H_{20}O$ menthol	hydroxyl group	alcohol		minty, medicinal
M	$C_{10}H_{18}O$ borneol	hydroxyl group	alcohol		camphor, pine

Questions

1. Circle and identify the functional group in each structural formula.
2. Can functional group alone be used to classify these five molecules according to their smell? Why or why not?

   No. These molecules each have a hydroxyl group and are alcohols, but three different smells are represented.

3. Is the molecular formula alone enough information to allow these five molecules to be classified according to their smell? Why or why not?

No. Each molecular formula can represent many different structures and molecules and more than one functional group.

4. Are there any structural similarities besides functional group that might be used to classify these molecules? If so, what are they?

Answers will vary. Similar shapes (for example, ring structures or the absence of ring structures, long molecules with two carbons branching off, or an extra "bridge" structure across the ring).

Part 2: Three-Dimensional Models

Examine the three ball-and-stick models. They represent the molecular compounds you smelled today. Figure out the molecular formula, name, and smell of each and write them in the table.

	Molecule 1	Molecule 2	Molecule 3
Molecular formula	$C_{10}H_{20}O$	$C_{10}H_{20}O$	$C_{10}H_{18}O$
Name	citronellol	menthol	fenchol
Smell	sweet (floral, lemon)	minty (medicinal)	camphor (pine)

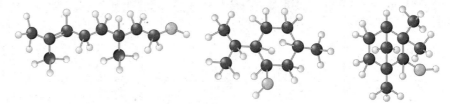

Questions

1. Compare the three models. List at least three physical differences that you notice between them.

Possible answers: The citronellol molecule does not have any ring structure. All its carbon atoms are in a jagged line. The fenchol molecule has a bridge across its ring structure. The citronellol molecule has a double bond in it.

2. **Making Sense** How is a ball-and-stick representation different from a structural formula? What additional information does it convey?

It is three-dimensional and shows how the atoms are arranged in space. It is tangible and can be taken apart and reconstructed. The atoms are different sizes and color-coded.

3. **If You Finish Early** Exactly what molecular model pieces would you need in order to construct a ball-and-stick molecular model of menthol?

You would need 10 black carbon atoms, 20 white hydrogen atoms, 1 red oxygen atom and 31 sticks for bonds.

Explain and Elaborate (15 minutes)

Discuss the Results of the Smelling

➡ Create a table on the board for vials I–M. Solicit students' smell categories. Guide them to the categories listed, and introduce the new smell classification "camphor."

Sample Questions

- What information did you discover about the molecules in vials I–M?
- There's a new smell category in two of the vials. How did you identify it?
- What similarities did you find among the structural formulas of the molecules that smelled similar?
- Are molecular formula and functional group enough information to predict smell? Why or why not?

Vial I	Vial J	Vial K	Vial L	Vial M
sweet	camphor	sweet	minty	camphor

Key Points

Each molecule in the activity has a hydroxyl group (−OH) and is an alcohol, but the molecules do not all smell the same.   Three distinctly different smells were present: sweet, minty, and a new smell we call "camphor." Camphor is a strong smell usually associated with mothballs and certain ointments. Apparently, something besides functional group is responsible for the smell of a compound.

There are a number of similarities among those alcohols that smell similar. The sweet-smelling alcohols in this group all have one oxygen atom and ten carbon atoms. In addition, the carbon atoms in the sweet-smelling alcohols do not form a ring. The camphor-smelling alcohols all have a ring structure with some sort of crosspiece or bridge. The minty-smelling alcohol has a ring but no bridge.

Discuss Ball-and-Stick Models

➡ Have the ball-and-stick models of citronellol, menthol, and fenchol at hand so you can discuss the various features.

Sample Questions

- What new information do you gain about molecules from looking at the ball-and-stick models?
- How is a ball-and-stick model similar to other representations you've seen?
- If you know the molecular formula of a molecule, can you build a ball-and-stick model of it? Why or why not?
- Describe the process you used to identify the three ball-and-stick models.
- If someone handed you a new model, do you think you could determine how the molecule it represents might smell just by looking at it?

Key Points

A ball-and-stick model shows the three-dimensional shape of a molecule.   The molecules are not flat, and the carbon atoms are not connected in a straight line.

Rather, the atoms are arranged in three dimensions in bent, crooked, or branched chains. The molecular formula and the structural formula of a molecule can both be determined from the ball-and-stick model. However, typically you cannot build a ball-and-stick model from just a molecular formula. The models in today's lesson are color-coded. All the hydrogen atoms are white; the carbon atoms are black, and the oxygen atoms are red. Sticks represent bonds between atoms, similar to the lines in structural formulas.

> **Ball-and-stick model:** A three-dimensional representation of a molecule that uses sticks to represent bonds and color-coded balls to represent atoms.

It is hard to tell at this point whether the shape of a molecule is related to its smell. There are some general structural similarities between the two sweet-smelling molecules and between the two camphor-smelling molecules. More information is needed to be able to generalize about molecular shape.

Wrap-up

Key Question: What three-dimensional features of a molecule are important in predicting smell?

- Molecular formula and functional group are not always sufficient information to predict the smell of a molecule accurately.
- A ball-and-stick model is a three-dimensional representation of a molecule that shows us how the atoms are arranged in space in relationship to one another.
- It appears that the smell of a compound may be related to its overall shape.

Evaluate (5 minutes)

> **Check-in**
>
> Predict the smells of these molecules.
>
> 1. Propyl butyrate
> 2. $C_6H_{14}O$
> 3. $H-\underset{\underset{H}{|}}{\overset{\overset{H}{|}}{C}}-\underset{\underset{H}{|}}{\overset{\overset{H}{|}}{C}}-\underset{\underset{H}{|}}{\overset{\overset{H}{|}}{C}}-\underset{\underset{H}{|}}{\overset{\overset{H}{|}}{C}}-\underset{\underset{H}{|}}{\overset{\overset{H}{|}}{C}}-\overset{\overset{O}{\|}}{C}-O-H$

Answers: 1. The first molecule, propyl butyrate, will probably smell sweet. You can tell it is an ester by its name, which ends in "-ate." So far we have lots of evidence that esters smell sweet. 2. The second molecule could smell several different ways because it could have any number of different structures. We've just seen that alcohols can be minty, sweet, or even camphor smelling. It could also be an ether. 3. The third molecule has a carboxylic acid functional group. It will probably smell putrid.

Homework

Assign the reading and exercises for Smells Lesson 9 in the student text.

LESSON 10
OVERVIEW

Two's Company
Electron Domains

Lesson Type
*Activity:
Pairs*

Key Ideas

Lone pairs of electrons have an effect on the shape of a molecule. The space occupied by a pair of electrons, whether a bonded pair or a lone pair, is called an electron domain. In a molecule, electron domains are located as far apart from one another as is physically possible. This distribution of electron domains results in the tetrahedral, pyramidal, and bent shapes of CH_4, NH_3, and H_2O, respectively.

As a result of this lesson, students will be able to

- determine the shapes of small molecules
- explain how lone pairs of electrons influence molecular shape
- describe electron domain theory and how it relates to molecular shape

Focus on Understanding

- Many textbooks refer to the ideas presented in this lesson as valence shell electron pair repulsion (VSEPR) theory. We have opted to refer to these ideas as electron domain theory instead, because the phrase "electron pair repulsion" can confuse students in light of the existence of double and triple bonds in molecules and in light of the confounding fact that two particles with the same charge should repel each other rather than pair up in a bond.

- Students might come away from this lesson thinking that all small molecules have an underlying tetrahedral shape; however, there are alternatives, such as BF_3, which is trigonal planar. These are covered in Lesson 11: Let's Build It.

Key Terms

electron domain
electron domain theory
tetrahedral shape
pyramidal shape
bent shape

What Takes Place

After being introduced to the concept of electron domains, students are provided with gumdrops, marshmallows, and toothpicks from which to create ball-and-stick models of methane, CH_4. They are challenged to manipulate and refine their model so that the electron domains end up as far apart as possible. This should result in a tetrahedral shape. When students successfully complete this step, they move on to build models of ammonia and water, with a student worksheet as a guide. Finally, groups of four create ball-and-stick models of four small molecules in order to further explore molecular geometry.

Materials

- student worksheet
- transparency—ChemCatalyst

Per pair

- gumdrops, 4 or 5 (slightly stale works best)
- mini-marshmallows, about 12 (slightly stale)
- toothpicks, about 15
- cm rulers

Per group of 4

- molecular modeling set or plastic bag containing 2 black spheres (4-hole), 2 red, 2 blue, 10 white, 6 paddles, 10 straight sticks

Setup

Use plastic sandwich bags as containers for sets of gumdrop, marshmallow, and toothpick materials for each pair of students. Slightly stale gumdrops and marshmallows work best and discourage consumption. Nevertheless, you might want to have surplus supplies on hand.

You could also bag up ball-and-stick model materials for the four models—methane, ammonia, water, and hydrogen fluoride.

Cleanup

The supplies can be saved for multiple class periods, although you will want to have surplus supplies to replace any that are damaged or lost. The candies and toothpicks are common household supplies and can be disposed of as such.

LESSON 10 GUIDE

Two's Company
Electron Domains

Engage (5 minutes)

Key Question: How do electrons affect the shape of a molecule?

ChemCatalyst

Examine the structural formula of ethanol. Which is the correct ball-and-stick model for ethanol? Explain your reasoning.

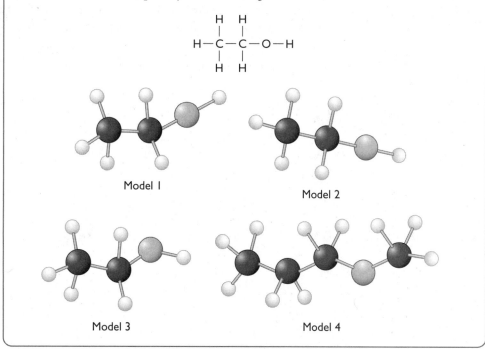

Sample Answer: Most students will easily eliminate model 4 because it has the wrong number of atoms. They may also eliminate model 2 because it is flat and not three-dimensional. However, they probably won't know whether model 1 or model 3 is correct. Listen to their reasoning. Model 3 is correct.

Discuss the ChemCatalyst

➡ Explore the subtleties of three-dimensional molecular shape.

Sample Questions

- Explain why each drawing is a correct or an incorrect model for ethanol.
- Why isn't model 2 correct? Model 4?
- Why do you think ball-and-stick models are crooked?
- How are models 1 and 3 different from each other?

Explore (20 minutes)

Introduce the Activity

➡ Go over the general instructions for building molecules from the materials provided.
- You will be using the gumdrops, marshmallows, and toothpicks as materials for ball-and-stick models.
- The toothpicks represent bonded electrons. The marshmallows represent hydrogen atoms. The gumdrops represent other atoms, such as carbon.
- Your challenge is to create a correct model for CH_4, methane. The only rule is that all the electron pairs must be equally distant from each other.
- Before you can continue, you must show your teacher this methane model.

Guide the Activity

➡ Challenge students to keep working if they show you incorrect methane models. If necessary, repeat the rule that all the electron pairs must be equally distant from each other. Point out which electrons in their model do not satisfy this rule.

➡ Students should not continue further on their worksheet until their tetrahedral methane model has been given the okay by you. Instruct students to continue to Part 2 when their model is complete.

➡ For Part 3, students can work in groups of four with one set of molecular models.

LESSON 10 ACTIVITY

Two's Company
Electron Domains

Name _____
Date _____ Period _____

Purpose
To use three-dimensional models to visualize small molecules.

Materials
- gumdrops, marshmallows, and toothpicks
- ruler
- ball-and-stick molecular model set

Part 1: Gumdrop Methane, CH_4

1. Create a methane molecule using gumdrops, marshmallows, and toothpicks.
2. Make sure every pair of electrons in the molecule is as far away as possible from every other pair of electrons. Use a ruler to check the distances.
3. Show your model to your teacher before proceeding.
4. Draw a picture of your final product.

Drawings will vary. A correct structure is shown here.

Part 2: Other Gumdrop Molecules

1. Draw Lewis dot structures for these molecules:

   a. CH_4 b. NH_3 c. H_2O

   H:C:H (with H above and below)   H:N:H (with H below, lone pair on top)   H:O: (with H below, two lone pairs)

2. How many pairs of electrons are located around the central atom of each molecule?

   Each central atom (C, N, and O) has four pairs of electrons around it.

3. Besides the identity of the central atom, what is different about these three molecules?

   Sample answers: Methane has four bonding pairs. Ammonia has three bonding pairs and one lone pair. Water has two bonding pairs and two lone pairs. Each molecule has a different number of H atoms.

4. Using gumdrops and toothpicks, create ball-and-stick models of NH_3 and H_2O.

5. Did you remember to include lone pairs in your models? How could you represent lone pairs?

   Answers may vary. You could use a toothpick to represent a lone pair.

6. If you need to, fix your models so that lone pairs are represented. Do the lone pairs have an effect on the shape of your molecule?

   Yes, because all pairs of electrons must be equally distant from each other.

7. Compare your three gumdrop models. Describe any similarities.

   Every molecule has essentially the same shape if the lone pairs are included.

Part 3: Ball-and-Stick Models

1. Work with your group. Use the ball-and-stick model set to create models of CH_4, NH_3, H_2O, and HF and draw them below. (Use black for carbon, white for hydrogen, and red for the other atoms.) Include the appropriate lone pair paddles in each model.

2. What do the models of these molecules have in common?

   With the lone pair paddles included, they all have the same basic tetrahedral shape and similar angles between atoms.

3. How many lone pair paddles would you need for an atom of neon? Explain your answer.

   Neon would need four lone pair paddles. It has eight valence electrons and four lone pairs.

4. What is the shape of each molecule if you ignore the lone pair paddles?

   Answers will vary. Something resembling the following: tetrahedral, pyramidal, bent, linear, point.

5. **Making Sense** Explain how the lone pairs affect the shapes of these molecules.

   Because all pairs of electrons must be as far from each other as possible, the lone pairs have as much effect on the shape as do the bonded pairs.

Explain and Elaborate (15 minutes)

Discuss the Gumdrop Model of CH₄

➡ Have on hand some of your own examples of incorrect models of CH_4 to discuss their relative merits and flaws.

Sample Questions

- Are all the C—H bonds identical in your model? How can you tell?
- If you were to measure the distances between each pair of atoms (or electron pairs) in your model, would they be the same? Are all the angles the same?
- The structural formula shows CH_4 as a two-dimensional cross. Are all the hydrogen atoms in a cross equidistant from one another?
- Show that each hydrogen atom in a tetrahedral molecule is equidistant from the other three hydrogen atoms.

Key Points

The overall geometric shape of a methane model is tetrahedral. In the tetrahedron, each bonding pair is equidistant from the other three pairs. All the H—C—H bond angles are identical (109°). The tetrahedral shape is identical any way it is turned in space.

In contrast, the bonding pairs are not equidistant from one another in a cross arrangement. In a cross, each bonding pair is 90° from two other bonding pairs but 180° from a third bonding pair. The bonding pairs can get farther apart from one another by moving out of the plane of the cross.

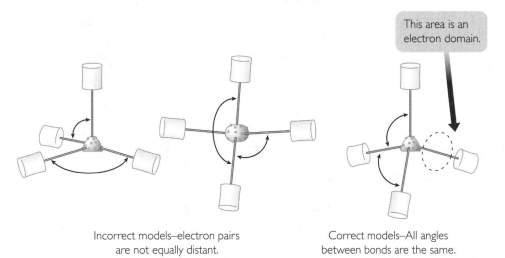

Incorrect models—electron pairs are not equally distant.

Correct models—All angles between bonds are the same.

This area is an electron domain.

> **Tetrahedral shape:** The shape around an atom with four bonded pairs of electrons. This is the shape of a methane molecule.

Introduce the Electron Domains

Key Point

An electron domain describes the area occupied by a set of electrons in a bond or a lone pair. Electron pairs tend to remain as far apart as possible from other pairs of electrons within a molecule. You can use the idea of electron domains to help determine the shape of a molecule.

> **Electron domain:** The space occupied by valence electrons in a molecule, either a bonded pair(s) or a lone pair. Electron domains affect the overall shape of a molecule.
>
> **Electron domain theory:** The idea that every electron domain in a molecule is as far as possible from every other electron domain in that molecule.

Discuss the Tetrahedral Shape of Molecules

⇒ Use the models of water and ammonia built by the students as examples.

⇒ At the appropriate time in the discussion, introduce the ball-and-stick models of methane, ammonia, water, and hydrogen fluoride.

⇒ Ask students to draw the structural formula of each molecule on the board.

Sample Questions

- How did the Lewis dot structures assist you with your gumdrop models?
- What gumdrop model did you create for NH_3? How did you arrive at that shape?
- What is the total number of electron domains in each of the three molecules in Part 2 of the worksheet?
- How do lone pairs influence the location of the hydrogen atoms?

Key Points

Even though the molecules you created today have different numbers of atoms, they all have a similar underlying shape. The underlying shape is called a tetrahedron. If you examine molecular models that include a special piece representing a lone pair (called a paddle), you can see the common geometries in these molecules (and even in a neon atom).

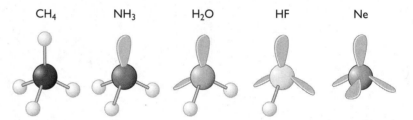

Most molecular models do not include lone pair paddles in their representations. Chemists simply remember that lone pairs are present in certain atoms. Lone pairs of electrons "take up space" and affect the ultimate shape of a molecule. The molecular model kits are designed so that the appropriate angles are created when the pieces are placed together. If you remove the lone pair paddles from the models, you see the more familiar representations of these molecules. In general, these shapes are referred to by the terms tetrahedral, pyramidal, bent, linear, and point. (Note: The pyramidal shape is trigonal pyramidal.)

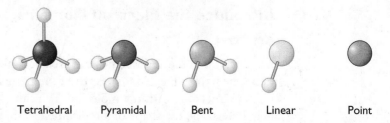

> **Pyramidal shape:** The shape around an atom with one lone pair of electrons. This is the shape of an ammonia molecule.
>
> **Bent shape:** The shape around an atom with two lone pairs of electrons. This is the characteristic shape of a water molecule.

Wrap-up

Key Question: How do electrons affect the shape of a molecule?

- Electron domains represent the space occupied by bonded electrons or a lone pair.
- Electron domains are located as far apart from one another as possible. This is referred to as electron domain theory.
- The three-dimensional shape of a molecule is determined by the valence electrons, both bonding electrons and lone pairs.

Evaluate (5 minutes)

> **Check-in**
>
> Use your model kit to build a model for ethanol. Be sure to use lone pairs to help you with your overall structure.
>
> ```
>       H   H
>       |   |
>   H — C — C — O — H
>       |   |
>       H   H
> ```

Answer: The model should look like model 3 in the ChemCatalyst. The geometry around each of the carbon atoms is tetrahedral. There are two lone pairs in this molecule, both on the oxygen atom. Because the lone pairs take up space, the C—O—H angle is less than 180°.

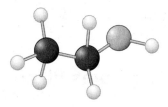

Homework

Assign the reading and exercises for Smells Lesson 10 in the student text.

LESSON 11

Let's Build It
Molecular Shape

OVERVIEW

Lesson Type
Activity:
Groups of 4

Key Ideas

Lewis dot structures are a useful tool in the three-dimensional modeling of molecules. They help us determine how many electron domains are in a molecule. Electron domain theory can be applied to molecules with double and triple bonds, resulting in other shapes besides tetrahedral. Although lone pairs influence the ultimate shape of the molecule, they are not included in the actual description of the shape of the molecule.

As a result of this lesson, students will be able to

- predict and explain molecular shape, including in molecules with multiple bonds

Focus on Understanding

- Lewis dot structures may lead students to think that bond angles must be 90° or 180°. Building ball-and-stick models gives students a better grasp of molecular shape.
- In the next lesson, smaller "shapes" come together to form larger molecules, describing another level of overall shape.

Key Terms

trigonal planar shape
linear shape

What Takes Place

Students use candy and toothpicks to build a model of formaldehyde, CH_2O, in order to learn about the effect of double bonds on the shape of a molecule. Groups of four then use ball-and-stick model-building materials to construct models of 13 different molecules, using Lewis dot structures as a guide. Students then draw and describe the different shapes of these molecules.

Materials

- student worksheet
- one ball-and-stick model of citronellol (from Lesson 9)

Per group of 4

- 2 gumdrops (preferably two different colors)
- 2 mini-marshmallows

- 4 toothpicks
- molecular modeling set, or plastic bag containing 2 black spheres (4-hole), 2 red, 2 blue, 6 white, 4 paddles, 6 straight sticks, and 3 curved sticks

Setup

Place any gumdrops, marshmallows, and toothpicks in bags or trays. Inventory the model-building sets. You may want to remove the space-filling modeling parts from the kit (disc connectors, white hemispheres, 3-hole carbons) so students have only ball-and-stick modeling parts to work with today.

Build one ball-and-stick model of citronellol for the Explain and Elaborate discussion.

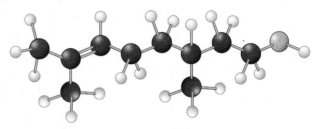

Citronellol

LESSON 11 — GUIDE

Let's Build It
Molecular Shape

Engage (5 minutes)

Key Question: How can you predict the shape of a molecule?

> **ChemCatalyst**
> 1. What is the Lewis dot structure of formaldehyde, CH_2O?
> 2. Draw formaldehyde's structural formula.
> 3. How many electron domains do you think this molecule has? Explain your reasoning.

Sample Answers:

1. [Lewis dot structure of formaldehyde, CH_2O] — There are three electron domains around the central carbon atom.

2. 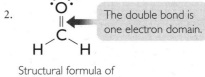 Structural formula of formaldehyde, CH_2O — The double bond is one electron domain.

3. There are three electron domains around the carbon atom: the two single bonds with hydrogen and the double bond with oxygen. Students might think that the double bond counts for two electron domains or that this molecule has five electron domains, because oxygen has two lone pairs. However, tell them you are most interested in the electron domains around the central carbon atom.

Discuss the ChemCatalyst

➡ Have students draw their Lewis dot structure and structural formula of formaldehyde on the board.

Sample Questions

- How many electron domains are there around the carbon atom? How do you account for the double bond?
- What overall shape do you think this molecule has? Is it tetrahedral?
- What effect does the double bond have on the overall shape of the molecule?

Explore (20 minutes)
Prepare for the Activity

➡ Pass out the gumdrop, marshmallow, and toothpick kits to groups of students to build a model of formaldehyde, CH_2O.

- Allow students to work out the shape of this molecule and describe it. The oxygen atom and hydrogen atoms need to be as far apart as possible.
- Tell students the shape of formaldehyde is referred to as trigonal planar. Point out that the overall shape is flat and triangular.

The shape of formaldehyde is trigonal planar

- Point out the double bond and the three electron domains around the central atom. Explain that the four electrons in the double bond are all in the same electron domain.
- Collect gumdrops and marshmallows before beginning the activity.

Introduce the Activity

- Pass out sets of ball-and-stick models. Tell students they will be drawing Lewis dot structures and using them as guides to create ball-and-stick models of the compounds in the table. Remind them that an electron pair is a bond shown with a line in a structural formula and a stick in a ball-and-stick model.
- Students can use black spheres for carbon, white spheres for hydrogen, red spheres for oxygen, and blue for nitrogen. Pairs of curved connectors can be used to make double-bonds.
- Advise students that some molecules may be large enough to be a combination of more than one shape. They should describe the shape around each central atom when necessary.
- Groups can divide up the work of building the models so that each group has one set of models to work with.

LESSON 11 ACTIVITY

Let's Build It!
Molecular Shape

Name _____
Date _____ Period _____

Purpose
To build ball-and-stick models from molecular formulas, using Lewis dot structures as a guide.

Materials
- ball-and-stick model-building materials

For each molecule, first draw the Lewis dot structure. Then build a three-dimensional molecular model, using the Lewis dot structures as a guide. Use black spheres to represent carbon, red for oxygen, white for hydrogen, and blue for other atoms.

Molecular formula	Lewis dot structure	Structural formula with lone pairs	Describe/draw shape
methane CH_4	H:C:H with H above and below	H–C–H with H above and below	tetrahedral
water H_2O	H:Ö: with H below	H–Ö: with H below	bent
ammonia NH_3	H:N:H with H below	H–N–H with H below	pyramidal
carbon dioxide CO_2	:Ö::C::Ö:	:Ö=C=Ö:	linear
chloromethane CH_3Cl	H:C:Cl: with H above and below	H–C–Cl with H above and below	tetrahedral
dichloromethane CH_2Cl_2	:Cl: H:C:Cl: with H below ; :Cl:C:Cl: with H above and below	H–C–Cl with Cl and H	tetrahedral

152 Unit 2 Smells
Lesson 11 • Worksheet

Living By Chemistry Teaching and Classroom Masters: Units 1–3
© 2010 Key Curriculum Press

Molecular formula	Lewis dot structure	Structural formula with lone pairs	Describe/draw shape
methanol CH_3OH	H:C:O:H with H's	H–C–O:H with H's	tetrahedral around the carbon atom, bent around the oxygen
ethane C_2H_6	H:C:C:H with H's	H–C–C–H with H's	tetrahedral around both carbon atoms
methyl amine CH_3NH_2	H:C:N:H with H's	H–C–N–H with H's	tetrahedral around the carbon atom, pyramidal around the nitrogen atom
formic acid CH_2O	:O: H:C::O:	:O: H–C=O:	trigonal planar around the carbon atom, bent around the oxygen atom
ethene (ethylene) C_2H_4	H:C::C:H with H's	H–C=C–H with H's	trigonal planar around both carbon atoms, flat molecule
hydrogen cyanide HCN	H:C:::N:	H–C≡N:	linear
ethyne (acetylene) C_2H_2	H:C:::C:H	H–C≡C–H	linear

Making Sense

How does the Lewis dot structure help you draw the structural formula?

Possible answer: they are virtually identical, except that electron pairs between atoms turn into lines representing bonds.

Explain and Elaborate (15 minutes)
Discuss the Model-Building Activity

- Ask students to share their models. Hold up each model in turn.
- Construct a table similar to the one below, containing the main shapes, as you solicit answers from students.

Number of domains around central atom	Number of lone pairs	Shape	Example	Sample sketch
4	0	tetrahedral	CH_4 (also CH_3Cl, CH_2Cl_2)	
4	1	pyramidal	NH_3	
4	2	bent	H_2O	
4 (around each carbon)	0	2 linked tetrahedrons	C_2H_6	
4 (around each central atom)	2 (on oxygen)	tetrahedral and bent	CH_4O	
2 (around the central atom)	2 (on each oxygen)	linear	CO_2	
3 (around carbon) and 4 (around oxygen)	2 (on each oxygen)	trigonal planar and bent	CH_2O_2	
3 (around each carbon atom)	0	2 linked trigonal planar (flat triangles)	C_2H_4	
2 (around carbon)	1 (on nitrogen)	linear	HCN (also C_2H_2)	

Sample Questions

- Identify the central atom or atoms in this molecule. How many electron domains are around the central atom?
- Which molecules have three electron domains around a central atom? What shapes do these molecules take? (trigonal planar) What shape do molecules with two electron domains take? (linear)
- What do double bonds do to the shape of a molecule?
- Both formaldehyde, CH_2O, and ammonia, NH_3, have four atoms. Why don't they have the same shape?

Key Points

Double or triple bonding changes the number of electron domains around an atom, affecting the overall shape of a molecule. In general, atoms tend to arrange themselves in tetrahedral shapes around atoms unless multiple bonds are present. When a carbon atom has a double bond and two single bonds, there are only three electron domains. The atoms arrange themselves in a flat triangle around the carbon atom. When a central atom has only two electron domains, as in carbon dioxide or hydrogen cyanide, the molecule ends up in a linear shape.

> **Trigonal planar shape:** A flat triangular shape found in small molecules with three electron domains surrounding the central atom.
>
> **Linear shape:** A geometric shape found in small molecules with two electron domains surrounding the central atom.

The number of electron domains is more important in determining the structure of a molecule than is the number of atoms. Both water and carbon dioxide have three atoms. However, water has a bent shape due to the four electron domains around the oxygen atom. Carbon dioxide, on the other hand, has two double bonds, resulting in a linear shape.

Discuss the Geometries of a Larger Molecule

➡ Use the ball-and-stick model of citronellol to show the overall effect of small shapes on the shape of a larger molecule.

Sample Questions

- What shapes are present within this large molecule of citronellol? (tetrahedral, trigonal planar, and bent)
- What causes shapes other than tetrahedral ones to show up in a molecule?
- Can you explain why the overall molecule is crooked instead of straight like the structural formula?

Citronellol

Section II Lesson 11 Let's Build It

Key Point

The more atoms in a molecule, the more combinations of shapes you might see together. Ethane has two carbon atoms. You can consider these both central atoms, each with four electron domains. The result is two tetrahedra stuck together at a vertex. A very large molecule, like citronellol, is simply a combination of many tetrahedra, with a trigonal planar area involving a double bond and a bent area next to the oxygen atom. The result is a molecule in which the electron domains are all as far apart as possible from one another.

In general, shapes other than tetrahedral ones show up in a molecule when there are multiple (double or triple) bonds.

Wrap-up

Key Question: How can you predict the shape of a molecule?

- Drawing the Lewis dot structure of a molecule allows us to predict its three-dimensional shape.
- The presence of double or triple bonds changes the number of electron domains around an atom, which in turn affects the overall shape of the molecule.
- The shape of large molecules is determined by the smaller shapes around individual atoms.

Evaluate (5 minutes)

> **Check-in**
>
> What is the shape of this molecule?
>
> H_2S

Answer: Bent, similar to water. Sulfur has six valence electrons, so it has two lone pairs, like oxygen. This gives sulfur four electron domains, which arrange themselves tetrahedrally. The result is a bent molecule.

Homework

Assign the reading and exercises for Smells Lesson 11 in the student text.

LESSON 12 OVERVIEW

What Shape Is That Smell?
Space-Filling Models

Lesson Type
Activity:
Groups of 4

Key Ideas

The overall shape of a molecular compound is directly related to its smell. Space-filling models provide another way to examine molecules three-dimensionally. Smell is related to the overall shape of the whole molecule.

As a result of this lesson, students will be able to

- build a space-filling molecular model given the structural formula
- begin to relate the overall shapes of molecules to their smell categories

Focus on Understanding

- Here we focus on the overall shapes of molecules, as opposed to the geometric shapes surrounding the central atoms, as we did in Lessons 10 and 11.
- We have carefully chosen examples to maintain simplicity and consistency in the smells categories. In reality, not every minty-smelling molecule is flat and planar, for example.

Key Term

space-filling model

What Takes Place

Topic: Molecular Modeling
Visit: www.SciLinks.org
Web code: KEY-212

Students use space-filling molecular models to examine the overall shape of six smell molecules. They compare and contrast similar-smelling compounds and different-smelling compounds, looking for shape trends. They find that there are three different overall shapes of these six molecules: stringy, ball-shaped, and frying-pan shaped. The Check-in involves an optional mini-drama in order to provide practice in hypothesis formation as well as to introduce a new smell molecule.

Materials

- student worksheet
- transparencies—ChemCatalyst and Check-in
- ball-and-stick model of citronellol for ChemCatalyst (optional)
- computer and projection system for viewing 3-D molecular models (optional)
- string tags or masking tape to label models with numbers 1–6
- mini-drama (optional): smock, goggles, gloves, tongs, empty sealed vial labeled vial Y, beaker, watch glass
- 1 or more sets of space-filling models of six smell molecules: two sweet (citronellol, geraniol), two minty (menthol, L-carvone), two camphor (camphor, fenchol) *or* 1 set of molecular model-building parts (optional)

Setup

To save class time, you could build all the space-filling models yourself ahead of time. Label them with the appropriate number (see worksheet). However, this is very time consuming—you will probably want help. Refer to the worksheet for the structural formulas. Here are the pieces you will need.

	4-hole carbons (black)	3-hole carbons (matte)	Oxygens (red)	Hydrogens (white hemispheres)	Single bonds (single disc)	Double bonds (double disc)
citronellol	8	2	1	20	9	1
geraniol	6	4	1	18	8	2
menthol	10	0	1	20	11	0
L-carvone	5	5	1	14	8	3
camphor	9	1	1	16	11	1
fenchol	10	0	1	18	12	0

If you have enough materials and want students to build the models during class, you might remove the ball-and-stick parts (long connectors, white spheres with holes) so that students have only the appropriate parts to build space-filling models today, and add "Build the space-filling models you have been assigned" to the start of the Instructions on the worksheet. Note that single-disc connectors represent single bonds, double-disc connectors represent double bonds, and hydrogen atoms need no connectors—they snap on directly. The 2-hole carbons (with matte finish) should be used for double-bonded carbons, with the double bond going into the "+" shaped hole.

If you decide to do the mini-drama, have the protective gear and vial Y hidden away but easily accessible.

Cleanup

Collect and inventory model-building sets or models.

LESSON 12 GUIDE

What Shape Is That Smell?
Space-Filling Models

Engage (5 minutes)

Key Question: How is the shape of a molecular compound related to its smell?

ChemCatalyst

What similarities and differences do you see between these two different types of models?

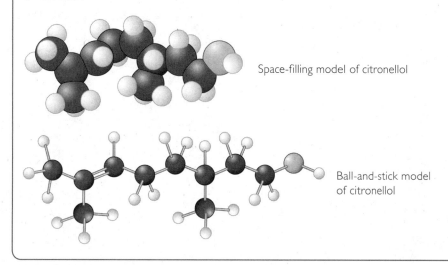

Space-filling model of citronellol

Ball-and-stick model of citronellol

Sample Answer: One model has sticks, and the other does not. One model is compact and chunky, and the spheres are partially hidden. Both models represent the same molecule, with the same number and color of each type of atom connected in the same order.

Discuss the ChemCatalyst

➡ Show 3-D computer images of molecules (optional)

Sample Questions

- How are the atoms represented in the space-filling model?
- How are the bonds represented in each model?
- Which model do you think is more accurate, or more like real life? What is your reasoning?
- Why do you think the atoms are not all perfectly round in a space-filling model?

Explore (15 minutes)

Introduce the Activity

- Tell students they will be examining another kind of three-dimensional model for molecules, called a space-filling model.
- Distribute a set of space-filling models or model-building sets to each group of four students. If students are building the models, assign each pair a different model to build. Then they can share their models to answer the worksheet questions.
- Go over what the pieces represent (see Setup).

LESSON 12 ACTIVITY
What Shape Is That Smell?
Space-Filling Models

Name _____
Date _____ Period _____

Purpose
To find connections between molecular shape and smell.

Materials
- set of space-filling models of six smell molecules

Instructions
You have space-filling models of six molecules. Match the space-filling models to the structural formulas here and on the next page. Put the number of the correct space-filling model in the corner of the box. (*Note:* Two boxes will be blank.)

1. Write the molecular formula of each compound next to its name in the table.

2. How are double bonds shown in a space-filling model? How can you figure out where a double bond is located in a space-filling model?

   Double bonds are not shown in most space-filling models. If a carbon atom is bonded to only three atoms, one of the bonds must be a double bond.

3. How is a space-filling model different from a ball-and-stick model? What new information does a space-filling model provide?

   Space-filling models show the overall shape of a molecule without the sticks of a ball-and-stick model. They better represent the shape of the molecule.

4. The following pairs of molecules smell similar. Sort the models into these three groups. Examine each pair of models and describe any structural similarities.

   citronellol and geraniol—smell sweet *Both are long strings of carbon atoms. Both have —OH and some branching.*

   menthol and carvone—smell minty *Both have a six-sided ring structure, branched extension, one oxygen atom and are kind of flat with a handle.*

   fenchol and camphor—smell medicinal *Both have a double-ring structure with five sides, one oxygen atom, and are rounded or balled up.*

5. What words best describe the overall shape of the three types of smells?

   sweet-smelling   minty-smelling   camphor-smelling
   long and crooked   *frying-pan*   *clump*

6. Predict the smells of isopentyl acetate and pulegone, located in the table. On what did you base your prediction?

   The isopentyl acetate should smell sweet. It has a structure similar to that of citronellol and geraniol, plus it is an ester. Pulegone should smell minty. It is shaped like the other two minty molecules.

154   Unit 2 Smells   *Living By Chemistry Teaching and Classroom Masters: Units 1–3*
Lesson 12 • Worksheet   © 2010 Key Curriculum Press

geraniol $C_{10}H_{18}O$

camphor $C_{10}H_{16}O$

citronellol $C_{10}H_{20}O$

6

5

1

isopentyl acetate $C_7H_{14}O_2$

fenchol $C_{10}H_{18}O$

pulegone $C_{10}H_{16}O$

2

L-carvone $C_{10}H_{14}O$

menthol $C_{10}H_{20}O$

4

3

7. Making Sense Do you think there is a connection between molecular shape and smell? Provide evidence to support your answer.

Quite possibly. For instance the two sweet-smelling molecules have different functional groups and different numbers of oxygen atoms, yet they both have a similar long, stringy shape.

Explain and Elaborate (15 minutes)

Assist Students in Sharing Their Observations

➟ Write the three smell classifications—sweet, minty, and camphor—on the board. At the appropriate time, make a list of some of the students' shape names under the smell headings.

Sample Questions

- Describe the process you used to identify the molecules represented by the models.
- Which do you think looks more like an actual molecule, a ball-and-stick model or a space-filling model? Explain your thinking.
- Which models represent similar smells? (1 and 6, 3 and 4, 2 and 5)
- What structural similarities did you find in the molecules that smell similar?
- How would you describe the overall shape of the sweet-smelling molecules? The minty-smelling molecules? The camphor-smelling molecules?

Key Points

A space-filling model is a three-dimensional model that a chemist uses to show how the atoms in a molecule are arranged in space and how they fill this space. Although we really can't say precisely what a molecule looks like, space-filling models could be considered slightly more accurate than ball-and-stick models. For one thing, covalent bonds between atoms cause atoms to be attracted to each other. Thus, atoms in a molecule probably are right next to each other rather than a stick's length apart.

In order to identify the space-filling models, you might have found it useful to start by figuring out the molecular formulas of the models and of the illustrations. You can also narrow down the answer by counting the number of oxygen atoms, looking for rings, or figuring out exactly where certain double bonds and oxygen atoms are located.

Molecules 1 and 6 both smell sweet. They also look similar to each other in overall shape—especially compared to the minty- and camphor-smelling molecules. The same can be said of the minty-smelling molecules and the camphor-smelling molecules.

Introduce Shape Names for Larger Molecules

➟ Starting from the students' suggestions, introduce the three shape names we will be using from now on.

Sample Questions

- What overall shape do the molecules in each smell category have?
- What did you predict for the smells of pulegone and isopentyl acetate? What was your reasoning?
- Do you have enough evidence to link smell to shape? (More data would be useful.)

Key Point

The shape of a molecular compound appears to be directly related to its smell. In order to talk about molecular shape, it is necessary to agree on some simple shape classifications. The sweet-smelling molecules in today's activity were all long and stringy molecules. The camphor-smelling molecules were also similar in shape—they looked bunched up. The minty-smelling molecules have a ring with a handle, sort of like a frying pan. For consistency we will call these shapes stringy, ball-shaped, and frying pan. We can summarize our observations:

Possible student answers:

Sweet	*Minty*	*Camphor*
long	ring-shaped	bunched up
stringy	flat	ball
snakelike	frying pan	cluster
stringy		ball-shaped

Note that these are hypotheses, not "rules" that apply to all molecules. While ball-shaped molecules typically are spherical, frying pan–shaped molecules do not always smell minty. Remind students that many more experiments are necessary to clarify the relationship between shape and smell.

Wrap-up

Key Question: How is the shape of a molecular compound related to its smell?

- Space-filling models provide another way of looking at the three-dimensional shape of molecules—one that represents the space occupied by atoms.
- Smell appears to be directly related to the three-dimensional molecular shape of a compound.
- The shapes of some large molecules can be described as stringy, ball-shaped, or frying pan.

Evaluate (10 minutes)

Complete a Mini-Drama Demonstration (optional)

➡ Dress up in protective garb and use protective equipment in order to dramatize the introduction of the molecule for the Check-in. Using tongs, remove the newest vial from a box and place it gingerly in a jar or beaker, and cover the container with something like a watch glass. Act as if you are handling a very dangerous substance.

➡ Write "Smell Y" on the board. Make sure the students see your actions, but don't give away anything.

➡ Students are likely to ask, "What are you doing?" or "Is that stuff dangerous?" or "Why are you wearing all that stuff?" or "What's in the bottle?" Point out that they are beginning an inquiry.

➡ Ask students what they think you are doing. They will say things like "We're smelling something dangerous," and "You're going to do a demonstration with a toxic substance." Point out that they have made a hypothesis about what you're doing. A hypothesis is an educated or best guess.

➡ Ask students how they arrived at their hypothesis and what evidence they have that supports their hypothesis. Suggest to students that they can make a hypothesis about the smell of the substance in vial Y. Tell them you are willing to give them some evidence in order to assist them in making their hypothesis. (Transition into the Check-in. If you want, reveal the information one piece at a time.)

T

> **Check-in**
>
> What smell do you predict for the substance in vial Y? Explain your reasoning.
>
> **Vial Y**
> Molecular formula: $C_{12}H_{20}O_2$
> Chemical name: bornyl acetate
>
> Structural formula   Molecular model

Answer: The two oxygen atoms indicate either a sweet or putrid smell. The chemical name indicates that it is an ester, with a sweet smell. The structural formula confirms the ester, but also has features of a minty- or a camphor-smelling molecule. The molecular model confirms a ball-shaped camphor-smelling molecule. In actuality, this molecule, bornyl acetate, is an insect pheromone. It smells both sweet and medicinal. It draws large numbers of insects to one place. (This is why we are not going to smell it and why the "mini-drama" emphasized caution!)

Homework

Assign the reading and exercises for Smells Lesson 12 in the student text.

LESSON 13

Sorting It Out
Shape and Smell

OVERVIEW

Key Ideas

Patterns in the chemical information of molecular compounds suggest general "rules" for some smell categories. Functional groups can be used to predict the smells of amines and carboxylic acids. Molecular shape is a predictor for minty-, sweet-, and camphor-smelling compounds. Even the chemical formula and name can be enough information to narrow down a smell.

Lesson Type
Activity:
Groups of 4

As a result of this lesson, students will be able to

- summarize the various connections explored so far between molecular structure and smell
- predict smells of a wide variety of compounds by examining molecular formulas, chemical names, molecular structures, and molecular shapes

What Takes Place

Groups of students are given a set of 24 molecule cards labeled A through X. The cards provide the chemical name, molecular formula, structural formula, and three-dimensional structure of these molecules. Students are challenged to sort the cards according to smell based on the evidence they've seen so far. Students use a worksheet to help them organize their discoveries. During the discussion, the class creates generalized rules for each smell category.

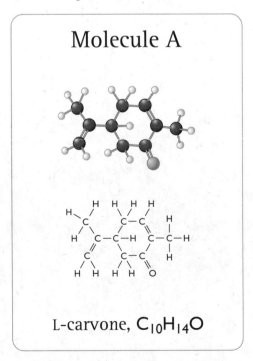

L-carvone, $C_{10}H_{14}O$

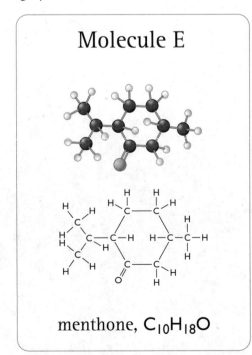

menthone, $C_{10}H_{18}O$

100   *Living By Chemistry Teacher Guide*   Unit 2 Smells

Materials

- student worksheet
- transparency—Smell Classification and Molecular Characteristics
- Smells Summary chart (from Lesson 6: Where's the Fun?)

Per group of 4

- set of 24 Molecule cards A–X

Setup

Inventory cards to make sure each set contains 24 different cards.

LESSON 13 GUIDE

Sorting It Out
Shape and Smell

Engage (5 minutes)

Key Question: What chemical information is most useful in predicting the smell of a compound?

> **ChemCatalyst**
>
> What smell or smells do you predict for a compound made of molecules that are long and stringy in shape? What is your reasoning?

Sample Answer: Most students would predict a sweet smell for a long, stringy molecule. All the evidence so far has pointed to this shape for sweet-smelling molecules.

Discuss the ChemCatalyst

➡ Review the evidence about smell.

Sample Questions

- On the basis of evidence collected in the previous class, how do you think a compound with stringy molecules would smell? Provide evidence to support your answer.
- What chemical information would you like to have in order to determine if a compound was made up of putrid or fishy-smelling molecules?

Explore (20 minutes)
Introduce the Activity

➡ Tell students they will work in groups of four to sort the cards into groups of molecules that will smell similar.

➡ Suggest to students that they put any cards they are unsure about into a separate pile.

LESSON 13 ACTIVITY

Sorting It Out
Shape and Smell

Name _____
Date _____ Period _____

Purpose
To help you determine which pieces of chemical information are most valuable in determining the smell of a compound.

Materials
- set of 24 Molecule cards

Instructions
Sort the cards as best you can into groups that will smell similar. Put any cards that your group is unsure about into a separate pile.

Questions

1. What features did you use to sort your cards?

   Answers will vary. We looked for the presence of functional groups, checked the chemical name, looked for the presence of nitrogen atoms, examined the overall shape of the molecules.

2. Which cards are you not 100% sure about?

   Answers will vary. However, many groups will mention molecules O, Q, and U and sometimes molecules R and W.

3. Explain what caused confusion about each card you were unsure how to sort.

   O—It is an alcohol. It is stringy, so it may smell sweet. Most sweet-smelling compounds are esters. Q—It is an alcohol. It is stringy, so it may smell sweet, but it is not an ester, like most sweet-smelling compounds. U—It looks like a ketone, with a carbon-oxygen double bond. It is stringy so it may smell sweet. It has a name that ends in "-al." R—It has the shape of a minty molecule; however, it is an alcohol. W—It has the shape of a minty compound, but it is a carboxylic acid, so it might be putrid.

4. Which group did you put these cards into? Explain why in each case.

   Card M: *Sweet. It has an ester functional group. It has two O atoms. Its name ends in "-ate," and it is long and stringy.*

   Card S: *Camphor. It has a ring with a bridge over it and is ball-shaped.*

   Card R: *Answers may vary. Minty, because it is frying pan–shaped. Or put it in the confused pile, because it is an alcohol and not a ketone like the other minty compounds.*

158 Unit 2 Smells
Lesson 13 • Worksheet

Living By Chemistry Teaching and Classroom Masters: Units 1–3
© 2010 Key Curriculum Press

5. Which (if any) compounds did you sort mostly by looking at shape?

   Answers may vary. The camphor, minty, and sometimes the sweet compounds can be sorted by shape.

6. Take the cards that your group is unsure about and sort them by shape. Where does each one end up? Is this correct?

   Answers may vary. O, Q, and U will end up in the sweet group, R in the minty group. W will end up in the minty group—this last one is incorrect. W is putrid.

7. What functional groups are present in the camphor-smelling compounds?

   ketone and alcohol

8. What functional groups are present in minty-smelling compounds?

   ketone and alcohol

9. Which smell classifications have only one functional group in them?

   fishy, putrid

10. Once you settle on your smell categories, fill in the following information for each smell. (In the third column, note whether the molecules have 1 O atom, 2 O atoms, etc.)

Smell classification	Cards	Molecular formula info	Shape(s)	Functional group(s)
sweet	C, D, F, H, L, M, O, Q, U	2 O atoms in some, 1 O atom in others	stringy	alcohol, ester, and aldehyde (Students will say it is a ketone.)
minty	A, E, J, R	rings, 1 O atom in all	frying pan	ketones and alcohols
camphor	P, S, T, V	double rings, 1 O atom in all	ball-shaped	ketones and alcohols
putrid	G, I, N, W	2 O atoms in all	stringy and frying pan	carboxyl
fishy	B, K, X	1 N atom in all	stringy and frying pan	amines

11. **Making Sense** How would you decide if a compound belongs in one of the five smell categories? What chemical characteristics would you look for first? Second?

    Answers will vary. Students might look for functional group first and sort out all the esters, amines, and carboxylic acids. They might double-check by looking at the names. Next they could use shape to sort out camphor, sweet, and minty.

Explain and Elaborate (15 minutes)

Discuss Student Observations

➡ Put the five smell classifications on the board. Ask groups to share their results.

➡ Reach consensus on the sorting of the cards. The molecule cards are placed in their correct smell classifications in the table below.

Sample Questions

- What process did you use to sort your cards?
- Which cards were you not 100% sure of? Why? Where did you decide to put them?
- What did this activity help you discover?

Generalize Characteristics for Determining Compounds' Smell Category

➡ Display the transparency Smell Classification and Molecular Characteristics. Fill in the table with class input. Circle the important characteristic for each smell category.

Smell classification	Cards	Molecular formula info	Shape(s)	Functional group(s)
sweet	C, D, F, H, L, M, O, Q, U	2 O atoms in some, 1 O atom in others	stringy	alcohol, ester, and aldehyde (Students should say it is a ketone.)
minty	A, E, J, R	rings, 1 O atom in all	frying pan	ketones and alcohols
camphor	P, S, T, V	double rings, 1 O atom in all	ball-shaped	ketones and alcohols
putrid	G, I, N, W	2 O atoms in all	stringy and frying pan	carboxyl
fishy	B, K, X	1 N atom in all	stringy and frying pan	amines

Sample Questions

- What do all the sweet-smelling compounds have in common? (stringy shape)
- What general statement can you make about carboxylic acids? (They are always putrid, no matter what their shape.)
- How can you tell a camphor-smelling compound from the others? (Its molecules will be ball-shaped.)
- What do you know about compounds that contain a nitrogen atom? (So far they are all fishy-smelling.)
- When is a ring-shaped compound not minty-smelling? (when it is an amine or a carboxylic acid)
- Are generalizations still useful even if we find some exceptions? Explain.

Key Point

In each smell category, it is possible to find one distinctive feature that sets that group apart from the other smell categories. For sweet, minty, and camphor compounds, the shape will always be consistent. Sweet compounds are stringy, minty compounds are frying-pan shaped with a six-carbon ring, and camphor compounds are ball-shaped. Putrid and fishy compounds, on the other hand, may take a variety of shapes. However, their functional groups are always consistent. So far in our investigation, all the putrid molecules have been carboxylic acids. All the fishy-smelling molecules have been amines.

Revise the Smells Summary Chart

➡ Tape up the Smells Summary chart you used in Lesson 6: Where's the Fun?. Have the students help you revise the information. (Changes are in bold.)

Sample Question

- How do we need to revise the Smells Summary Chart to bring it up to date?

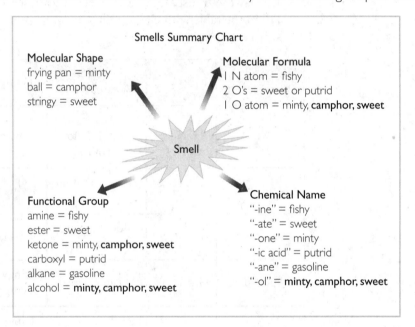

Wrap-up

Key Question: What chemical information is most useful in predicting the smell of a compound?

- Molecular shape can be useful in predicting smells for sweet-, minty-, and camphor-smelling compounds.
- Amines and carboxylic acids have distinctive smells.
- For stringy and frying-pan shaped compounds, it is necessary to look at functional group as well as molecular shape in order to determine smell.

Evaluate (5 minutes)

> **Check-in**
>
> If a compound is sweet smelling, what other things do you know about it? List at least three things that are probably true.

Answer: A compound that is sweet smelling probably contains two oxygen atoms and an ester functional group. One of the oxygen atoms is double-bonded. The chemical name of the molecule probably ends in "-ate." It has stringy molecules.

Homework

Assign the reading and exercises for Smells Lesson 13 in the student text.

LESSON	How Does the
14	Nose Know?
OVERVIEW	Receptor Site Theory

Key Ideas

Lesson Type
Activity:
Groups of 4

Scientists have come up with many models to explain the chemistry of smell. Most agree that the shape of a molecular compound is important to the way the compound smells. The receptor site model is one of the most widely accepted models of smell. In this model, molecules with different shapes are like keys that fit into locks, or receptors, in the nose. Molecules with different shapes trigger different smells.

As a result of this lesson, students will be able to

- come up with a plausible model to explain how smell works in the nose, based on the evidence thus far
- describe the receptor site model

Focus on Understanding

- The receptor site model, sometimes referred to as "lock-and-key," is considered one of the primary ways molecules trigger responses in the body. For example, pain medication is detected in receptor sites. Thus, it is an extremely useful concept for students.
- During the presentations, classmates may ask clarifying questions but should not attempt to either disprove or validate anyone's model.
- Students might incorrectly assume that molecules break apart when they become airborne. The fact that molecular shape is important in determining smell provides evidence that molecules actually stay together. This concept is fleshed out in the reading for this lesson.

Key Term

receptor site theory

What Takes Place

Groups of students speculate as to what takes place in the interior of the nose in order for us to detect the different smells of different compounds. Each group comes up with a model to explain the mechanics of smell and creates a poster showing how the model works. Each group then presents its poster to the class, explaining the intricacies of the model. After the presentation, students are introduced to the receptor site model. *Note:* This lesson may take more than one standard class period. If your time is limited, direct students to keep their presentations to one minute, or select only a few groups to present.

Materials

- transparency—Check-in
- handout—How Does the Nose Know?

Per group of 4

- poster paper
- at least 4 felt-tipped marking pens in various colors

Setup

Make sure you have some way to post or display the students' posters during the groups' presentations.

LESSON 14 GUIDE

How Does the Nose Know?
Receptor Site Theory

Engage (5 minutes)

Key Question: How does the nose detect and identify different smells?

ChemCatalyst

1. Suppose you needed to separate coins but could not see them. Explain how you would make a machine that detects and sorts coins.
2. How do you think your nose detects a smell?

Sample Answers: 1. Students might suggest sorting them by size. If you shake the coins over a series of holes, those that are smaller than the hole will fall through. 2. Some students might have heard of the receptor site model and may say that molecules fit into spaces in the nose or that the nose can detect the shapes of functional groups, or that the molecules travel up into the brain where they are detected.

Discuss the ChemCatalyst

➡ Discuss students' ideas about how the nose works.

Sample Questions

- What is a smell?
- Can you only smell things that are in the air? Explain.
- What do you think is going on in the nose when you smell something?
- When you have a cold, you can't smell things as well. What do you think is happening that affects your sense of smell?

Explore (20 minutes)

Introduce the Activity

➡ Give students the handout for today's activity.

➡ Divide students into groups of four. Tell them they will be designing a model for how smell works and illustrating it on a poster. Their mechanism must explain how the smell reaches the nose, what happens to smell molecules when they are in the nose, and how we perceive smell and identify it with other experiences.

➡ Let students know they will present their models to the class.

➡ Tell students they will have limited time to complete their posters (perhaps ten minutes of group discussion and ten minutes of poster creation).

➡ Let students know when half the time has passed so they can get started drawing if they are not already doing so.

LESSON 14 ACTIVITY

How Does the Nose Know?
Receptor Site Theory

Name _____
Date _____ Period _____

Purpose
To build a model to explain how your nose detects different smells.

Materials
- poster paper
- different-colored markers

Instructions

1. Today you will be designing a model for smell and illustrating how it works on a poster. Your model should
   - Explain how a smell reaches the nose.
   - Explain the chemistry involved in smelling something.
   - Explain how we perceive or interpret the smell in the brain and relate it to other experiences.
2. Once you have finished your model, give it a descriptive name, for example, the molecular mass model.
3. Summarize your model in one or two sentences at the bottom of your poster.
4. Be prepared to present your model to the class. For example, be prepared to describe how your nose is able to detect the difference between the two molecules shown here.

Ball-shaped molecule

Stringy molecule

Explain and Elaborate (20 minutes)
Group Presentations of Smell Models

➡ Allow a minute or two for each presentation. You might choose one student from each group to be the spokesperson. Provide positive feedback for ideas.

➡ Check whether the model proposed by each group explains the evidence we have about the sense of smell.

Evidence about the sense of smell:

- You can detect a smell at a distance from the source. Wafting works.
- The shape of the molecule is related to how it smells.
- Some people cannot detect specific smells.
- You cannot detect smells when you have a cold.
- The sense of smell degrades as you get older.
- We all react similarly to certain bad smells like rotten eggs.
- Smells are connected with memories.
- For sweet, minty, and camphor molecules, the shape will always be consistent. Sweet molecules are stringy, minty molecules are frying-pan shaped, and camphor molecules are ball-shaped.
- Putrid and fishy molecules, on the other hand, may take a variety of shapes. However, their functional groups are always consistent.
- So far in our investigation, all the putrid molecules have been carboxylic acids, and vice versa. All the fishy-smelling molecules have been amines, and vice versa.

➡ Get the class to assist in making suggestions for improvement of each model to support the evidence. (*Note:* Remind students that this is what scientists do. They test hypotheses and try to improve their models based on evidence. Remind the class that scientists to this day do not know the details of how the sense of smell works.)

Sample Questions

- How do molecules get from a substance into your nose?
- How does the nose distinguish between molecules with different shapes and functional groups?
- What evidence do you have that molecules do not fall apart when traveling through the air?
- What evidence is there that our brain is involved?
- What experiments could you do to test each model?

Key Points

A good model should show that the nose can distinguish between molecules with different shapes and functional groups.   Students might propose that molecules with the same functional groups react chemically inside the nose or that molecules with similar shapes fit into receptor sites. They might suggest that nose hairs collect smells. A good model should explain why some people cannot detect certain smells and why you cannot smell as well when you have a cold.

A model might show that there is a connection to the brain.   This explains why memories are associated with smell. It can also account for the deterioration in the sense of smell as you get older. The connection to the brain may also be the reason we react similarly to bad smells such as rotten eggs and to good smells such as flowers.

A model might demonstrate that molecules remain intact. We have found that the shape of a molecule of a compound is important to the way it smells. This means that the molecule must enter the nose intact, not as a cluster of atoms.

Introduce Current Receptor Site Theories (optional)

➡ Point out any student models that happen to match the scientists' models.
➡ Highlight the receptor site model. Draw several rudimentary receptor sites on the board to demonstrate.

Key Points

Scientists have proposed many theories about how smell works and created models corresponding to these theories. The last model described here (model 4) is called the receptor site model; it reflects the most widely accepted theory about how the interior of the nose detects different smells.

Model 1. Molecules vibrate inside the nose, with each distinct molecule vibrating differently. Our nerves and brain determine smell by detecting these different molecular vibrations.
Model 2. Smell molecules react chemically with the inside of the nose in such a way that our nerves detect and our brain deduces smells from the chemical reactions.
Model 3. Molecules of compounds press against the inside of the nose or even punch through it. This causes chemical or physical changes inside the nose that allow us to determine the smell of the compounds.
Model 4. Smell molecules of compounds fit into receptor sites, which are like molds that mirror the shape and size of the molecules. Each receptor site is specific to one shape. When there is a molecule sitting in a particular receptor site, it generates a signal to our brain and we register this as a specific smell.

> **Receptor site theory:** The currently accepted model explaining how smells are detected in the nose. Molecules fit into receptor sites that correspond to the overall shape of the molecule. This stimulates a response in the body.

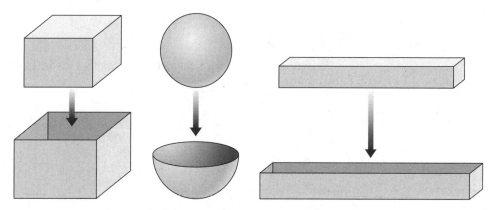

Some overly simplified receptor sites and the shapes that might fit into them.

Wrap-up

Key Question: How does the nose detect and identify different smells?

- The currently accepted model for smell describes smell molecules landing in receptor sites that fit the shape of the smell molecules.
- In the receptor site model, each receptor site has a specific shape that corresponds to the shape of just a few smell molecules.

Evaluate (5 minutes)

Check-in

One of the molecules that gives coffee its smell is 2-furylmethanethiol.

1. Write down everything you know about how this molecule is detected by the nose.
2. Draw a possible receptor site for this molecule.

Answers: 1. We know that a molecule must first enter the nose. It must be airborne, and it must be inhaled. It lands on the lining of the nose and fits into a specific receptor site that is shaped similarly to the molecule. The sulfur atom in this molecule may also affect its smell.
2. A receptor site for this molecule might look something like a pentagon with a bump on it.

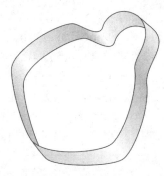

Homework

Assign the reading and exercises for Smells Lesson 14 and the Section II Summary in the student text.

SECTION III

Molecules in Action

Section III focuses on interactions between molecules. Lesson 15 is a charged wand demonstration and lab that provide evidence of polarity in molecules. Lesson 16 uses an information-packed comic strip to develop the concept of polarity and dipoles. Lesson 17 introduces the electronegativity scale. Students use electronegativity values to compare atoms and bonds and to figure out the direction of bond dipoles. In Lesson 18, students go one step further, using electronegativity values to determine the direction of polarity of entire molecules. They discover that perfectly symmetrical molecules are nonpolar and often do not smell. Lesson 19 explores how molecules, polarity, phase, shape, and bonding patterns are related to smell.

In this section, students will learn

- the role of polarity in smell chemistry
- about electronegativity and bond dipoles
- about the continuum of bonding from nonpolar covalent bonds to ionic bonds
- to predict whether a substance will have a smell based on its composition, bonding, phase, size, and polarity
- how polarity relates to intermolecular interactions

LESSON 15 OVERVIEW

Attractive Molecules
Attractions Between Molecules

Lesson Type
Lab and demo:
Groups of 4

Key Ideas

Molecules have an equal number of protons and electrons and no net charge. However, many molecules have regions of positive and negative partial charge. These molecules are called polar molecules. Polarity in molecules contributes to intermolecular attractions. Polarity affects the overall behavior of a substance and may play a role in smell properties. Water is one of the best-known polar compounds.

As a result of this lesson, students will be able to

- describe the behavior of polar molecules
- explain the general difference between a polar and a nonpolar molecule
- describe basic intermolecular attractions
- define a partial charge

Focus on Understanding

- Students might not realize that "static" is really static electricity and occurs when an object has gained or lost some electrons. Depending on the rod and the fabric used, the rod may develop a positive or a negative charge, but it will attract the polar liquids either way.

- Water is a small polar molecule, yet for humans it does not have a smell. Some students might notice this discrepancy. The subject is handled in Lesson 18. Put simply, there are so many H_2O molecules in the nose that we do not smell the "additional" water molecules.

- Students might think of a force as something imposed on an object from an outside source. Emphasize the "attraction" aspect of intermolecular forces: The molecules are attracting each other.

Key Terms

polar molecules
nonpolar molecules
intermolecular forces

What Takes Place

Students first consider why air does not smell. Then the instructor demonstrates the attraction of a stream of water to a charged wand. Students consider possible interactions between water molecules and the charged wand and then work in groups of four trying the charged wand on three new liquids located at different stations, all but one of which are attracted to the wand. Students also examine a

116 *Living By Chemistry Teacher Guide*   Unit 2 Smells

droplet of each of the four liquids placed on wax paper: Three bead up, but one doesn't. Polarity is introduced, and intermolecular interactions are put forth as a possible factor in the smells of compounds. Optional: You could leave out a set of drops overnight and have students observe the rate of evaporation, another indication of the intermolecular forces present.

Materials

- student worksheet
- transparencies—ChemCatalyst, Charged Wand, and Water Molecules Interacting
- 6 burets and buret stands
- 3 polar liquids (e.g., water, vinegar, rubbing alcohol), 200 mL each
- 1 nonpolar liquid (e.g., hexane), 200 mL
- 500 mL beakers (6) to hold liquid
- 6 droppers for liquids
- charged wand sets (plastic or Lucite rod with wool or silk cloth)
- waxed paper

Setup

Before class, set up burets containing three different liquids at stations around the room (one or more stations per liquid, depending on your class size). Set up a 500 mL beaker under each buret to catch the stream of liquid. A wand and a soft cloth should also be placed at each station. Include an index card or sticker with the name of the liquid. Also have pieces of waxed paper and droppers available at each station, or set up another station with the same liquids and waxed paper. Set up one buret at the front of the room for a demonstration with water.

Cleanup

Store liquids for use with future classes.

LESSON 15 GUIDE

Attractive Molecules
Attractions Between Molecules

Engage (5 minutes)

Key Question: Why do some molecules smell while others do not?

ChemCatalyst

If a molecule fits into a receptor site in the nose, it seems as if it should smell. Yet most of the molecules in air—O_2 (oxygen), N_2 (nitrogen), CO_2 (carbon dioxide), and Ar (argon)—do not have a smell. What do you think is going on?

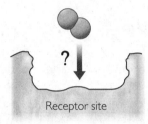

Receptor site

Answers: Students might say that small molecules simply "fall out" of receptor sites or are too small to fit the site or that air smells but we are just used to it so we don't notice it. They might also say that the smell of air is too faint for humans to detect. All are reasonable answers.

Discuss the ChemCatalyst

➡ Stimulate discussion about why some substances don't smell.

Sample Questions

- Why do you think clean air doesn't smell?
- If the small molecules in air fit into receptor sites, why wouldn't they smell?
- What do you think it would be like if we could smell the air around us all the time?
- Do you think molecules in the air interact with one another? Explain.
- Do you think molecules in the air interact with the nose? Explain.

Explore (20 minutes)

Prepare for the Lab

➡ Demonstrate the effect of a charged wand on water. Use the charged wand and a fine flow of water from a buret to provide macroscopic evidence that some molecules are attracted to a charge.

- Rub a plastic wand with a cloth to create a charge on the wand.
- Hold the charged wand close to, but not touching, the stream of liquid.
- Repeat, moving the wand close to and away from the stream of water.

➡ Use the demonstration as an opportunity to introduce students to the use of a buret.

Introduce the Lab

➡ Tell students that they will be using this same technique to test three more liquids in class. The liquid should not touch the charged wand.

➡ Direct students how to rotate through the stations.

➡ Tell students that if the burets are close to empty they should be refilled carefully with the contents of the catch beaker. The valve on the buret *should be closed* during refilling.

LESSON 15 LAB

Attractive Molecules
Attractions Between Molecules

Name _____
Date _____ Period _____

Purpose
To observe the response of certain liquids to an electrical charge and the behavior of the same liquids as droplets.

Part 1: Testing the Liquids
With your group, test the liquid at each station with a charged wand. Next place a drop of the liquid on waxed paper. Enter the results in the table.

Compound	Effect of charged wand	Behavior on waxed paper
water	attracts	round drop
acetic acid	attracts	round drop
isopropanol	attracts	round drop
hexane	no response	spreads out flat

Part 2: Analysis

1. Here is an artist's interpretation of what is happening between the water molecules and the charged wand. Write a paragraph describing in your own words what you think is happening in the picture.

   Answers will vary. Hopefully, students will notice that the water molecules have pluses and minuses on them. In addition, the charged wand is shown with negatively charged electrons on it. The illustration shows how the positively charged portion of each water molecule is arranged so that it faces the negatively charged wand. Thus, the water is attracted to the wand because opposite charges attract. This explains why the stream of water bends toward the wand.

2. What evidence do you have that some of the molecules you just tested may have a charge on them?

Three of the liquids moved closer to the charged wand when it was brought near. They also beaded up on the waxed paper.

3. How do you explain any liquids that are not attracted to the charged wand?

Answers will vary. Speculation by students is okay. The hexane did not move toward the charged wand. It doesn't have a charge on its molecules.

4. How is the behavior of the droplets related to the charged wand experiment?

The liquids that bead up are also those that are attracted to a charged wand. These liquids may be attracted to each other, causing them to bunch together or bead up.

5. **Making Sense** If water molecules are carrying a partial charge, as shown, how do you think a group of water molecules would behave toward each other? To illustrate your thinking, draw a picture of several water molecules interacting. Explain your drawing.

A single water molecule

Answers will vary. Hopefully, students will draw a picture showing the water molecules interacting with one another in some appropriate way.

164 Unit 2 Smells
Lesson 15 • Worksheet

Explain and Elaborate (15 minutes)
Discuss the Results of the Charged Wand Activity

→ Display the transparency of water interacting with the charged wand.
→ Draw a large water molecule, with its partial charges, on the board, or refer to the illustration on the transparency. Introduce the delta symbol.

Sample Questions

- What did you observe when you tested the various liquids with the charged wand?
- What do you think is going on at a molecular level to attract the liquids to the wand?
- Why do you think the charged wand does not affect some liquids?
- Do you think the same liquids would be attracted to a positive charge? Why or why not?

Key Points

The charged wand experiment provides evidence that some molecules are attracted to a charge. The most obvious explanation for this observation is that those molecules must have some sort of charge on them. This is true. Some molecules have a slight charge on opposite ends of the molecule. These molecules are called polar molecules. According to our results today, water, acetic acid, and isopropanol are all polar molecules. Other molecules, such as hexane, are *not* attracted to the charged wand. Such molecules are called nonpolar molecules.

> **Polar molecules:** Molecules that are attracted to a charge because they have partial charges on them.
> **Nonpolar molecules:** Molecules that are not attracted to a charge.

One end of a polar molecule has a partial negative charge, and the other end of the molecule has a partial positive charge. It is important to stress that this charge is "partial," as compared to the type of charge on an ion. In order to differentiate between full charges and partial charges, chemists use the symbols δ+ and δ− (delta plus and delta minus) to indicate partial charges. The hydrogen atoms in a water molecule have partial positive charges, while the oxygen atom has a partial negative charge (as shown in the drawing).

The individual molecules in polar liquids will respond when another charged substance comes near. In the illustration, you can see that the water molecules are all pointed in the same direction. This is because the positively charged ends of the water molecules are attracted to the negatively charged wand. The overall attraction between the water and the wand causes the liquid to move in the direction of the wand.

Discuss Intermolecular Interactions

- You might want to have students draw their interpretations of the Making Sense question on the board. Ask them to explain their drawings.
- For the first key point, display the transparency that shows water molecules interacting.

Sample Questions

- What did you observe during the droplet exercise?
- Explain what is happening in your Making Sense drawing.
- How are the water molecules interacting on the transparency? Did any of your drawings come close to matching this illustration?
- What effect do you think these interactions might have on the properties of water?
- Explain how molecular interactions might account for the beading behavior observed in some of the liquid droplets.

Key Points

The partial charges on polar molecules cause individual molecules to be attracted to each other. As the molecules in a polar liquid tumble around, they tend to align with each other such that the side that is partially negative is closer to the partially positive side of another molecule. The attraction that happens between individual polar molecules is called an **intermolecular force** or an intermolecular attraction. (The prefix *inter* means "between.")

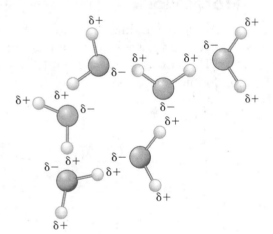

> **Intermolecular forces:** The forces of attraction that occur between molecules.

These interactions are responsible for many observable properties of polar substances. For instance, we observed in this lesson that polar liquids tend to bead up because the individual molecules are attracted to each other and "cling together."

All molecules interact with each other, but the attractions between polar molecules tend to be stronger than those between nonpolar molecules. In general, polar molecules are more likely to be liquids rather than gases at room temperature. This is because the attractions between individual molecules cause them to form a liquid rather than to disperse as a gas.

Section III   Lesson 15 Attractive Molecules   123

However, even with nonpolar molecules, the random motion of electrons can cause momentary imbalances of charge, resulting in a momentary partial charge. This force, called the London dispersion force, draws even nonpolar molecules together to some extent.

Wrap-up

Key Question: Why do some molecules smell while others do not?

- Polar molecules have partial charges on parts of the molecule.
- Polar molecules are attracted to a charge.
- Polar molecules are attracted to each other. These intermolecular interactions account for many observable properties.

Evaluate (5 minutes)

> **Check-in**
>
> Acetone is polar. Name two other things that are probably true about acetone.

Answer: Because acetone is polar, it probably is a liquid at room temperature. In addition, it should bead up when placed on a piece of waxed paper. Students might also say that acetone will have a smell and that it is attracted to a charged wand.

Homework

Assign the reading and exercises for Smells Lesson 15 in the student text. Exercise 4 is a lab report for today's lab. Optional: Assign students to read "The Bare Essentials of Polarity" comic strip in the student text in preparation for the next lesson.

LESSON 16 OVERVIEW

Polar Bears and Penguins
Electronegativity and Polarity

Lesson Type
Activity:
Pairs

Key Ideas

The tendency of an atom to attract shared electrons is called electronegativity. When two different atoms bond together, the atom that is more electronegative will attract the shared electrons, causing a partial negative charge on that atom. A molecule is nonpolar if it has no partial charges or if the partial charges cancel. A molecule is polar if it has a positive end and a negative end. There is a continuum of bonding from equal sharing to unequal sharing of shared pairs of electrons due to differences in electronegativity. Electronegativity is covered more quantitatively in the next lesson.

As a result of this lesson, students will be able to

- explain what causes polarity and polar molecules
- describe the different types of bonding that correspond to different combinations of electronegative atoms
- predict the general direction and strength of a dipole for any two atoms, using the periodic table

Focus on Understanding

- The term *electronegative* can seem contrary to some students. It is not the tendency of an atom to be negative. Rather, it is the tendency of an atom to attract negatively charged electrons.
- The word *dipole* can be a source of confusion because it has several only subtly different meanings. Chemists refer to polar molecules as dipoles, and they also say that individual bonds have dipoles (which are numeric and measurable). Finally, molecules with polar bonds can have net dipoles, also called a dipole moment.

Key Terms

dipole   electronegativity

What Takes Place

Students read the multipage comic strip, "The Bare Essentials of Polarity," which focuses on polarity, electronegativity, and bonding. They answer questions on a worksheet to analyze the illustrations in the comic strip. The concept of electronegativity is introduced, along with three different categories of bonds.

Materials

- student worksheet
- handout—"The Bare Essentials of Polarity" (also available in the student text)
- transparency—ChemCatalyst

LESSON 16 GUIDE

Polar Bears and Penguins
Electronegativity and Polarity

Engage (5 minutes)

Key Question: What makes a molecule polar?

ChemCatalyst

Consider this illustration:

1. If the penguin represents a hydrogen atom and the polar bear represents a chlorine atom, what does the ice cream represent in the drawing? What do you think the picture is trying to illustrate?
2. Would HCl be attracted to a charged wand? Explain your thinking.

Sample Answers: 1. Most students will figure out that the ice cream represents the electrons shared between the atoms. Other answers are possible. The picture is trying to illustrate that the chlorine atom is somehow stronger than the hydrogen atom and pulls on the electrons more.
2. Students might say that HCl will be attracted to a charged wand because it has a lopsided molecule that is more negative on one side than on the other. Some might say they cannot tell from the evidence.

Discuss the ChemCatalyst

➡ You might have a student draw the Lewis dot structure of HCl on the board to remind students of the shared electrons.

Sample Questions
- What do you think the polar bear and penguin drawing is trying to illustrate?
- Why is the penguin being swept off his feet by the polar bear?
- How successfully are the polar bear and penguin sharing the ice cream cone?
- What explanation of polarity does the ChemCatalyst illustration provide?

Explore (20 minutes)
Introduce the Activity

⟹ Have students individually read the multipage comic strip, "The Bare Essentials of Polarity," before passing out the worksheets. Students can work in pairs on the worksheet.

LESSON 16 ACTIVITY

Polar Bears and Penguins
Electronegativity and Polarity

Name _____
Date _____ Period _____

Purpose
To understand polarity and bonding between atoms.

Instructions
Read the comic strip "The Bare Essentials of Polarity," and use it to answer these questions.

1. How does the comic strip define a polar molecule?

   A polar molecule is a molecule with a difference in electrical charge between two ends.

2. Define electronegativity as you understand it, after reading the first two pages of the comic strip.

   Answers will vary according to the individual student's understanding. Sample answer: Electronegativity is the tendency of an atom or nucleus to attract bonding electrons toward itself.

3. What is the artist trying to represent by two polar bears arm wrestling or two penguins arm wrestling?

   The artist is trying to show two identical atoms wrestling for the shared electrons in the bond. Their sizes and strengths are identical, so no animal is shown beating the other. The arm wrestling ends in a tie.

4. What three types of bonds are represented on the third page of the comic strip? What happens to the bonding electrons in each type of bond?

   Nonpolar covalent bonds: Equal sharing of the bonding electrons.

   Polar covalent bonds: The bonded electrons spend more time around the more electronegative atom. This results in a polar molecule with partial charges.

   Ionic bonds: A bonding electron (or electrons) is given up to the more electronegative atom. This leaves ions with full positive and negative charges.

5. Explain why there are four scoops of ice cream in the illustration of O_2 on the third page.

   A double bond is represented. There are four shared electrons.

6. What do the six scoops of ice cream represent in the illustration of N_2 on the fourth page?

   They represent six shared electrons, or a triple bond between two nitrogen atoms.

7. Describe what you think is happening to the penguin in the CO_2 molecule in the picture on the fourth page.

   The penguin is being pulled equally in both directions by the more electronegative polar bears. The bears will get more of the ice cream.

8. Name three things that the picture of CO_2 on the fourth page illustrates about the molecule.

   The molecule is nonpolar, because the equal pulls on the electrons cancel each other out. The oxygen atoms are more electronegative than the carbon atom. The bonding electrons will spend more time around the oxygen atoms than around the carbon atoms. The carbon dioxide molecule is linear.

9. Describe what you think is happening to the penguins in the illustration of H_2O on the fourth page.

   The penguins are being dragged away by the polar bear, but they are still strong enough to hang onto the ice cream cones. The bear represents the more electronegative oxygen atom pulling on the electrons shared with each hydrogen penguin. The net result of these two polar bonds is a dipole on the entire molecule.

10. What does the crossed arrow represent in the comic strip?

    It represents a dipole. It points in the direction of the more electronegative atom. It also shows which end of the molecule is positive, by the crossed end.

11. What are two of the definitions of dipole given in the comic strip?

    A dipole can be a polar bond, or it can be a polar molecule. Also, the actual net polarity of a molecule sometimes is referred to as a dipole.

12. **Making Sense** What does electronegativity have to do with polarity?

    Answers will vary. Electronegativity essentially "causes" polarity. When one atom attracts electrons more than the other atom in a bond, it results in the electrons hanging around that atom. This causes a partial negative charge on the more electronegative atom and a partial positive charge on the less electronegative atom.

13. **If You Finish Early** Using polar bears and penguins, create an illustration showing a hydrogen sulfide molecule, H_2S. (*Hint:* You might want to start with a Lewis dot structure.)

    Answers will vary. Hopefully, students will create a drawing similar to the one for water, H_2O, because hydrogen sulfide has a similar shape.

You don't have to go to the ends of the earth to find polar molecules. They're all over the place. A polar molecule is just a molecule with a difference in electrical charge between two ends.

Polarity in molecules is caused by differences in electronegativity between atoms. Electronegativity describes the ability of an atom to attract bonding electrons toward itself.

Chlorine is more electronegative than hydrogen. So the bonded pair of electrons in HCl spends more time near chlorine.

Electronegativity values tend to increase as you move "northeast" on the periodic table, and decrease as you move "southwest."

The noble gases are often not assigned electronegativity values. They rarely bond to other atoms.

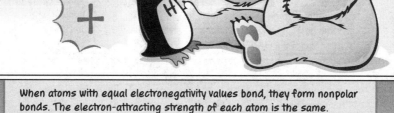

Because the elements have such varying electronegativities and can bond in many different combinations, there is really a continuum of polarity in bonding. We can break the continuum down into three categories.

NONPOLAR COVALENT

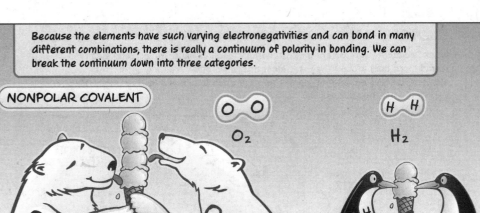

The clearest examples of nonpolar covalent bonds are those between identical atoms, such as in H_2, N_2, O_2, or Cl_2. Bonds between atoms with nearly the same electronegativity value, such as carbon and hydrogen, can also be considered nonpolar.

POLAR COVALENT

In a polar covalent bond, two atoms share bonded pairs of electrons somewhat unequally. The electrons are more attracted to one atom than the other. Examples include bonds between carbon and oxygen atoms, or between hydrogen and fluorine atoms.

IONIC

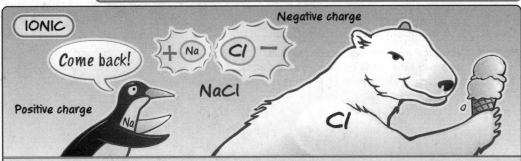

A large difference in electronegativity results in the winner-take-all situation of ionic bonding. The more electronegative atom takes the bonding electrons and becomes a negative ion, while the other atom becomes a positive ion. The opposite charges on the ions attract each other.

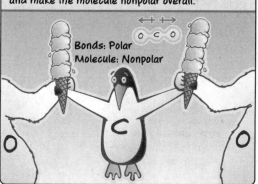

Explain and Elaborate (15 minutes)

Define Electronegativity

⇒ Draw a molecule of HCl on the board. At the appropriate time in the discussion, add the partial charges and the dipole.

Sample Questions

- List some of the things you learned about polarity from the comic strip.
- How would you use polar bears and penguins to illustrate a polar molecule? A nonpolar molecule?
- According to the comic strip, which elements tend to attract shared electrons to the greatest degree?
- According to the comic strip, what is electronegativity?

Key Point

The tendency of an atom to attract shared electrons is called electronegativity. When two atoms with different electronegativities bond, they attract the bonding electrons to different degrees. The bonding electrons spend more time around one of the atoms, resulting in a partial negative charge on that atom. An atom that strongly attracts the shared electrons is considered highly electronegative. The atom with lower electronegativity will end up with a partial positive charge on it. The result is a polar bond, and the polar molecule is a dipole. Hydrogen chloride, HCl, is shown here as an example.

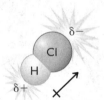

Hydrogen chloride, HCl
Electrons pulled in the direction of the dipole arrow.

> **Electronegativity:** The tendency of an atom to attract the electrons that are involved in bonding.
>
> **Dipole:** A polar molecule or a polar bond between atoms. A crossed arrow is used to show the direction of a dipole. The crossed end of the arrow indicates the partial positive (+) end of the polar bond, and the arrow points in the direction of the partial negative (−) end.

Relate Electronegativity to Bonding

⇒ Provide students with a general description of the three categories of bonds summarized at the end of the comic strip.

Sample Questions

- How does the comic strip illustrate the range of possible ways of sharing electrons?

- What are the names of the types of bonds that form between atoms? How are these bonds different from one another?
- Why do some elements in molecules have partial charges?
- What is an ion? How does an ion form?

Key Point

Bonds that involve sharing or transferring electrons fall into three categories. In nonpolar covalent bonds, the electrons are shared equally (e.g., in diatomic molecules such as H_2, the atoms are identical, so there is no difference in the degree to which each atom in the molecule attracts the shared electrons). In polar covalent bonds, the electrons are shared unequally. The electrons are more attracted to atoms with higher electronegativity. Ionic bonds involve the transfer of one or more electrons from one atom to another. When two atoms have *very* different electronegativities, the bond between them is considered ionic. The dividing line between polar covalent bonding and ionic bonding is not clear-cut.

Wrap-up

Key Question: What makes a molecule polar?

- Polarity in a molecule is caused by unequal sharing of electrons between atoms.
- Electronegativity is the tendency of an atom to attract shared electrons.
- Anytime two atoms with different electronegativity values share electrons, there will be a partial negative charge on one atom and a partial positive charge on the other atom.
- Bonds are classified as nonpolar covalent, polar covalent, and ionic as the difference in electronegativity between the two atoms in the bond increases.

Evaluate (5 minutes)

> **Check-in**
>
> Consider hydrogen iodide, HI.
>
> 1. Is HI a polar molecule? Explain your reasoning.
> 2. How would the atoms be portrayed in the comic strip—as polar bears, penguins, or both? Explain.

Answers: 1. Hydrogen iodide is a polar molecule. Its atoms do not share the electrons equally. Iodine is located on the right side of the periodic table, where the electronegativities are greater. Hydrogen is on the left side. 2. In a drawing, hydrogen would be a penguin and iodine would be a polar bear. The iodine atom would have a partial negative charge on it.

Homework

Assign the reading and exercises for Smells Lesson 16 in the student text.

LESSON 17 OVERVIEW

Thinking (Electro)Negatively
Electronegativity Scale

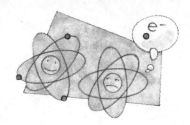

Lesson Type
Classwork: Individuals

Key Ideas

The electronegativity scale is an arbitrary scale invented by Linus Pauling in 1932 that assigns a numerical value to the electronegativity of each atom in the periodic table. By comparing the electronegativities of two atoms, chemists can determine whether a resulting bond between them will be highly polar. The numbers that Pauling generated allow us to quantify the continuum of bonding—nonpolar covalent, polar covalent, and ionic bonds.

As a result of this lesson, students will be able to

- use the electronegativity scale to compare atoms and to compare (calculate) the polarity of different bonds
- use the electronegativity scale to predict bond dipoles and bond type
- describe the continuum of nonpolar, polar, and ionic bonding in terms of electronegativity

Topic: Electronegativity
Visit: www.SciLinks.org
Web code: KEY-217

Focus on Understanding

- While the noble gases are considered unreactive, there are occasional exceptions and sometimes they are assigned electronegativity values. To keep from confusing students, we have not included this information.
- It can be difficult for students to grasp the idea of a unitless scale.

Key Term

diatomic molecule

What Takes Place

The concept of electronegativity that was introduced in the polarity comic strip in Lesson 16 is expanded on. Students learn to read and use the electronegativity scale. Students use electronegativity values to compare atoms and bonds and to figure out the direction of bond dipoles. They also apply electronegativity values to the continuum of bonding from nonpolar covalent to polar covalent to ionic.

Materials

- student worksheet
- transparencies—ChemCatalyst, Electronegativity Scale, and Bonding Continuum
- handouts—Electronegativity Scale and Bonding Continuum

LESSON 17 GUIDE

Thinking (Electro)Negatively
Electronegativity Scale

Engage (5 minutes)

Key Question: How can electronegativity be used to compare bonds?

ChemCatalyst

1. Explain how the illustration and the table below it might be related to each other.
2. What patterns do you see in the numbers in the table?

Sample Answers: 1. Some students will say that the numbers in the table must stand for the electronegativity of each atom. The sizes of the polar bears and penguins represent the numbers. The thickness of the ice also is related to the magnitude of the numbers. 2. The numbers in the table increase across a row and decrease down a group.

Discuss the ChemCatalyst

Sample Questions

- How is the illustration related to the table shown below it?
- What do you think the numbers on the table represent?
- What parts of the table do the penguins and polar bears represent? What does the ice represent?
- What do the polar bears and penguins look like where the elements are most chemically reactive?
- Which animals represent nonmetals? Why are there some bears in the metals area of the periodic table?

Explore (15 minutes)

Introduce the Classwork

- Tell students they will be exploring electronegativity in a quantitative way.
- Pass out the worksheets and the handout Electronegativity Scale. Students can work individually or in pairs.

LESSON 17 CLASSWORK
Thinking (Electro)Negatively
Electronegativity Scale

Name _____
Date _____ Period _____

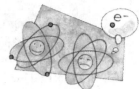

Purpose
To explore numerical values for electronegativity and to learn how to use them to compare atoms and bonds.

Questions
Use the handout Electronegativity Scale to answer these questions.

1. What happens to the electronegativity values across each period from left to right?

   Electronegativity values generally increase from left to right across the periodic table.

2. What happens to the electronegativity values of each group from bottom to top?

   Electronegativity values generally increase from the bottom to the top of the periodic table.

3. If you have a bond between a metal atom and a nonmetal atom, which of the two is more electronegative? Explain your thinking.

   The nonmetal atom generally is more electronegative. It is located on the right side of the periodic table, where the more electronegative atoms are located.

4. Where are the atoms with the greatest electronegativity values located? Are they metals, nonmetals, or metalloids?

   nonmetals, located in the upper right corner of the periodic table

5. Where are the atoms with the lowest electronegativity values located? Are they metals, nonmetals, or metalloids?

   metals, located in the lower left corner of the periodic table

6. Metals often are referred to as electropositive. Explain why.

   Metals have low electronegativities, which means that the shared electron pair spends more time around the other atom in the bonded pair. Hence, metals tend to have partial positive charges.

7. Why do you think the noble gases do not have electronegativity values?

   Electronegativities refer to a pair of bonded electrons, and noble gases usually do not form bonds. Thus, electronegativities are not defined for noble gases.

8. Circle the atom in each pair below that will attract shared electrons more strongly.
   a. C or (Cl) **b.** Rb or (Br) **c.** (I) or In **d.** Ag or (S) **e.** (As) or Na **f.** H or (Se)

9. Which two atoms in the periodic table form the most polar bond?

   fluorine and francium, F and Fr

10. List at least three examples of pairs of atoms with nonpolar covalent bonds.

    any two atoms of the same element—for example, H—H, O=O, F—F

11. If the difference in electronegativity between two bonded atoms is greater than 2.1, then the bond is considered ionic. List three examples of pairs of atoms with ionic bonds.

    alkali metals and alkaline earth metals bonded to oxygen or to the halogens—for example, Na—Cl, K—Br, Mg—O

12. If the difference in electronegativity between two bonded atoms is less than 2.1, then the bond is considered polar covalent. List three examples of pairs of atoms with polar covalent bonds.

    hydrogen and metals on the right side of the periodic table bonded with nonmetals—for example, O—H, Zn—S, Ga—As

13. Metal atoms tend to form cations with positive charges. Is this consistent with the electronegativity of metal atoms? Why or why not?

    Yes; metal atoms tend to transfer electrons rather than attract electrons. In the case of an ionic bond, the electrons are transferred to the nonmetal.

14. Provide two explanations for why nonmetals tend to form anions.

    Nonmetals are among the most electronegative elements, so they tend to attract electrons to form anions with negative charges. Nonmetals can gain one or more electrons to achieve the same electron configuration as the noble gas at the end of the row.

15. Sulfur forms both ZnS and SF_2. Is sulfur the most electronegative element in both compounds? Why or why not?

    In ZnS, the S atom is the more electronegative of the two atoms, so it has a partial negative charge. In SF_2, the F atom is more electronegative, so the S atom has a partial positive charge.

16. Arrange these bonded pairs in order of increasing polarity from the least polar to the most polar: C—H, H—O, N—H, and H—F.

    C—H < N—H < H—O < H—F

17. **Making Sense** Explain how you would use the electronegativity scale to determine both the direction and the degree of polarity of a bond between two different atoms.

    Find the electronegativity of each atom from the table. Then determine the difference between the electronegativities of the atoms. Large differences indicate a very polar bond. The atom with the larger electronegativity is the negative side.

18. **If You Finish Early** Which of these pairs of atoms would result in the most electronegative bond? The least electronegative bond? Arrange them in order from the least polar to the most polar.

    C—H   C—S   H—F   C—N   C—O   H—Br

    C—S < C—H < C—N < C—O < H—Br < H—F

Explain and Elaborate (20 minutes)

Discuss the Table of Electronegativities

→ Display the transparency Electronegativity Scale.

Sample Questions

- What general trends in electronegativity do you notice from the table?
- Why are metals referred to as electropositive?

Key Point

In 1932, Linus Pauling created a scale for electronegativity and assigned numerical values for the electronegativities of the elements. The scale ranges from 4.0 down to 0, with 4.0 the highest electronegativity. The lowest electronegativities can be found in the lower-left portion of the periodic table and in the noble gas family, and the highest electronegativities are found in the upper-right portion of the table. The scale was invented by comparing the polarities of a variety of molecules. This scale is unitless.

Electronegativity scale

H 2.10																	He
Li 0.98	Be 1.57											B 2.04	C 2.55	N 3.04	O 3.44	F 3.98	Ne
Na 0.93	Mg 1.31											Al 1.61	Si 1.90	P 2.19	S 2.58	Cl 3.16	Ar
K 0.82	Ca 1.00	Sc 1.36	Ti 1.54	V 1.63	Cr 1.66	Mn 1.55	Fe 1.83	Co 1.88	Ni 1.91	Cu 1.90	Zn 1.65	Ga 1.81	Ge 2.01	As 2.19	Se 2.55	Br 2.96	Kr
Rb 0.82	Sr 0.95	Y 1.22	Zr 1.33	Nb 1.60	Mo 2.16	Tc 1.90	Ru 2.2	Rh 2.28	Pd 2.20	Ag 1.93	Cd 1.69	In 1.78	Sn 1.96	Sb 2.05	Te 2.1	I 2.66	Xe
Cs 0.79	Ba 0.89	La* 1.10	Hf 1.30	Ta 1.50	W 2.36	Re 1.90	Os 2.20	Ir 2.20	Pt 2.28	Au 2.54	Hg 2.00	Tl 1.62	Pb 2.33	Bi 2.02	Po 2.00	At 2.20	Rn
Fr 0.70	Ra 0.89	Ac* 1.10															

Discuss Bonding Between Pairs of Atoms

→ Display the transparency Bonding Continuum.

Sample Questions

- Explain how you would determine which end of a molecule with two atoms has a partial negative charge and which end has a partial positive charge.
- Which pairs of atoms do you expect to be most polar?
- When two atoms have a very large difference in electronegativity, the bond between them is considered ionic. Explain why.
- What is a polar covalent bond?
- Where would Cl_2 be on the continuum of bonding? (nonpolar covalent) HCl? (polar covalent)

Key Points

By determining the numerical difference between electronegativities in a bond, you can compare the polarities of bonds. For instance, hydrogen has an electronegativity of 2.10. Chlorine has an electronegativity of 3.16. The difference in their electronegativities is 1.06, making HCl a polar molecule.

Numerical differences in electronegativity can also help predict the type of bond that will be found.

No difference in electronegativity: When the two atoms bonded together are identical, the electrons are shared equally. The bond is called nonpolar covalent.

Small differences in electronegativity (below 2.1): When there is a small difference in electronegativity between two atoms bonded together, there is unequal sharing of electrons. The electrons are attracted toward the atom with the greater electronegativity. Partial charges are set up on each atom, and the bond is called polar covalent.

Large differences in electronegativity (above 2.1): When there is a large difference in electronegativity between two atoms bonded together, the electron essentially is transferred from one atom to the other. The more electronegative atom gets the electron(s) and becomes a negative ion. The bond is called ionic, because ions are formed.

Note: When the difference in electronegativity between two atoms is very small (less than 0.5), the bond is so slightly polar that it is often considered nonpolar. For example, the electronegativity difference between carbon and hydrogen is less than 0.5, so it is often considered nonpolar.

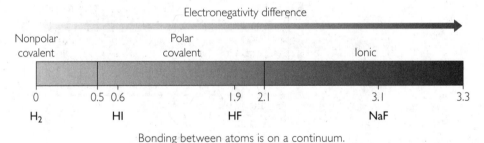

Bonding between atoms is on a continuum.

The dividing line between polar covalent bonding and ionic bonding is not clear-cut. The value 2.1 given for the difference in electronegativity as the dividing line is only a guide. Bonding between atoms is on a continuum.

Wrap-up

Key Question: How can electronegativity be used to compare bonds?

- Electronegativity measures how strongly an atom will attract shared electrons.
- The greater the difference in electronegativity between two atoms, the more polar the bond will be.
- In ionic bonding, the electronegativities between two atoms are so different that we can think about the bond as one in which the electron(s) of one atom is (are) completely transferred to the other atom.

Evaluate (5 minutes)

> **Check-in**
> 1. Is the bond in potassium chloride, KCl, nonpolar, polar, or ionic? Explain.
> 2. To what degree do the K and Cl atoms in KCl, potassium chloride, share electrons?

Answers: 1. The electronegativity of K is 0.82, and the electronegativity of Cl is 3.16. The difference is 2.34. In fact, the difference in electronegativity is so great that we can think about the bond as one in which one electron is given up by the potassium atom and is transferred to the chlorine atom. The bond is ionic. 2. The potassium and chlorine atoms do not really share the electron.

Homework

Assign the reading and exercises for Smells Lesson 17 in the student text.

LESSON 18 OVERVIEW

I Can Relate
Polar Molecules and Smell

Key Ideas

The shape and symmetry of a molecule can affect its polarity. In general, polar molecules smell, and nonpolar molecules do not smell. Polarity affects smell in two ways. First, it affects whether a molecule will dissolve in the watery mucous lining of the nose. Second, the fact that polar molecules are attracted to each other may assist them in docking with the polar portions of the large protein molecules that make up the receptor sites.

Lesson Type
Activity: Pairs

As a result of this lesson, students will be able to

- assess a molecule for symmetry and determine whether it is likely to be polar
- use electronegativity values to locate the partial negative and partial positive portions of a molecule
- explain the connection between polarity and smell

Focus on Understanding

- Students often assume that the location of a lone pair will be the location of a partial negative charge. This is not always the case.

What Takes Place

Using the electronegativity scale, students figure out the location of partial charges on several molecules. This assists them with determining the direction of polarity. Students are given paper representations of eight molecules that they must cut out and orient properly in relationship to a receptor site and a molecule of water. Students are challenged to determine which molecules do not smell, and why. Students learn about the importance of polarity in smell chemistry and are introduced to current theories about receptor sites, polarity, and smell.

Materials

- student worksheet
- handout—Molecules Cutouts
- ball-and-stick models (optional)
- handout—Electronegativity Scale (from Lesson 17, Thinking (Electro)Negatively)

Per pair
- scissors
- glue stick

Setup

You might want to have a few ball-and-stick models available at the front of the class to demonstrate the symmetry of small nonpolar molecules and the asymmetry of small polar molecules. Suggested nonpolar models: CF_4, CO_2, N_2, CH_4. Suggested polar molecules: CF_3Cl, H_2O. You might also want to have a medium-size molecule such as citronellol available.

LESSON 18
GUIDE

I Can Relate
Polar Molecules and Smell

Engage (5 minutes)

Key Question: What does polarity have to do with smell?

> **ChemCatalyst**
>
> Hydrogen chloride, HCl, and ammonia, NH_3, have a smell, and large amounts of each dissolve in water. Oxygen, O_2, nitrogen, N_2, and methane, CH_4, do not have a smell, and only a small amount of each dissolves in water. How can you explain these differences?

Sample Answer: Students might say that the molecules that dissolve in water are more polar than those that don't dissolve much in water. They may or may not add that polarity must have something to do with smell.

Discuss the ChemCatalyst

➡ You might want to pursue the ChemCatalyst question in depth by having students draw structural formulas (including lone pairs) of these molecules on the board.

Sample Questions

- How are the molecules different? (HCl and NH_3 are asymmetrical, polar molecules, whereas O_2, N_2, and CH_4 are symmetrical, nonpolar molecules)
- How are the molecules in the ChemCatalyst similar? (They are all relatively small.)
- Why do you think HCl and NH_3 dissolve easily in water? Why do you think they have a smell?
- How do you think polarity is related to smell?

Explore (20 minutes)
Introduce the Activity

➡ Students can work in pairs. Pass out the worksheet, the handout, glue sticks, and scissors. To save time and cutting, you could have each pair share a worksheet.

➡ Students will also need the handout Electronegativity Scale from Lesson 17: Thinking (Electro)Negatively.

LESSON 18 ACTIVITY

I Can Relate
Polar Molecules and Smell

Name _____
Date _____ Period _____

Purpose
To practice determining the polarity of molecules with more than two atoms and to relate polarity to smell.

Materials
- handouts—Molecules Cutouts, Electronegativity Scale, scissors, glue stick

Instructions
1. Cut out the paper molecules provided for you on the handout. Put the water molecules aside for now.
2. Use the electronegativity scale to determine the location of partial negative and partial positive charges on each molecule.
3. Label the partial charges on the molecules using δ+ and δ− signs.
4. Imagine that these drawings represent receptor sites with a partial negative charge. Figure out how each cut-out molecule would align itself in relationship to the receptor site. Then glue it in place.

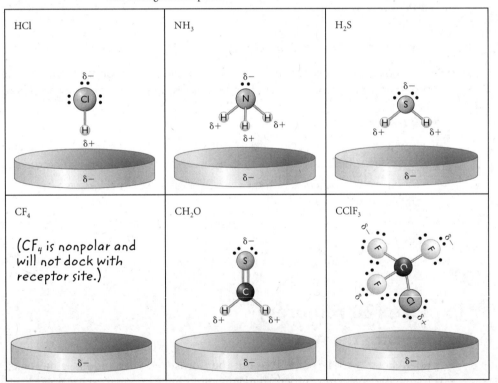

| HCl | NH$_3$ | H$_2$S |
| CF$_4$ (CF$_4$ is nonpolar and will not dock with receptor site.) | CH$_2$O | CClF$_3$ |

180 Unit 2 Smells
Lesson 18 • Worksheet

Living By Chemistry Teaching and Classroom Masters: Units 1–3
© 2010 Key Curriculum Press

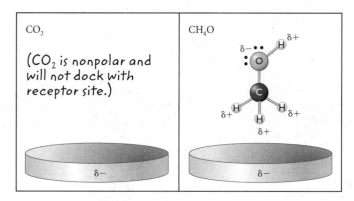

5. How are $CClF_3$ and CF_4 similar? How are they different?

   Both are tetrahedral. Both have a central carbon atom with four additional atoms. They are shaped like methane. Both have 12 lone pairs. However, the molecule with the chlorine atom is polar, while the other molecule is nonpolar.

6. Two of the molecules you cut out do not have a smell. Which ones do you think they are, and why?

   Carbon tetrafluoride, CF_4, and carbon dioxide, CO_2, don't have a smell. They are both nonpolar molecules because of their overall shape and symmetry.

7. The chart shows some molecules that have a smell and some that have no smell. What do you think those that have a smell have in common? What about those that don't have a smell?

   The molecules that have a smell are all polar molecules. Those that don't have a smell are all nonpolar. The nonpolar molecules are also very symmetrical.

8. Ammonia, NH_3, dissolves in water. Use the four water molecules you have cut out to determine how an ammonia molecule might interact with water. Be prepared to demonstrate your answer to the class.

   To be discussed in class.

9. Is water a polar molecule? Why can't you smell it?

   Students will be speculating on the answer. Water is a polar molecule. Maybe we can't smell it because it is so prevalent in the nose already.

Molecule	Has a smell?
N_2	no
PH_3	yes
CH_4	no
H_2Se	yes
NH_3	yes
HBr	yes
CO_2	no
AsH_3	yes

10. **Making Sense** How can you identify which molecules are nonpolar? Explain the process you would use.

    First, look for symmetry in overall shape. If there is no symmetry, then the molecule is polar. If there is symmetry, check to see whether or not the electronegativities of the atoms cancel each other out.

MOLECULES CUTOUTS

Cut out these molecules on the dashed lines.

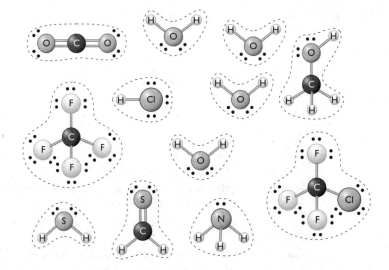

Explain and Elaborate (15 minutes)

Discuss the Polarity of Molecules with More Than Two Atoms

- Have your own set of cut-out molecules available for demonstration. You might want to use tape to stick them on the board or on a piece of butcher paper.
- Ask students to come up and draw the location and identity of the partial charges on each molecule.

Sample Questions

- What process did you use to figure out the polarity of a molecule?
- Do the lone pairs tell you where the more electronegative part of the molecule is? Explain.
- What is the best way to determine which part of a molecule has a partial negative charge and which part has a partial positive charge?

Key Points

Polarity of diatomic molecules is fairly easy to determine. If the two atoms are identical (such as two hydrogen atoms), then the molecule is not polar. If the two atoms are different (such as HCl), then you must figure out which atom is more electronegative. That is the end with the partial negative charge.

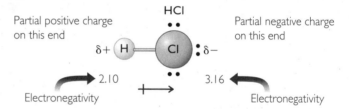

Both electronegativity and the overall symmetry of the molecules help determine the polarity of molecules with more than two atoms. For example, to figure out the polarity of formaldehyde, CH_2O, all you need to know is that oxygen is more electronegative than hydrogen. Likewise, the nitrogen atom in ammonia, NH_3, is more electronegative than the hydrogen atoms. If you can't remember electronegativity trends, you can always check the electronegativity chart.

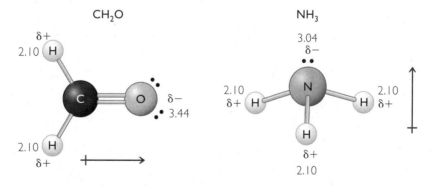

Relate Polarity to Smell

- You might want to have ball-and-stick models on hand in order to demonstrate the symmetry of small nonpolar molecules. Suggested models: CF_4, CO_2, N_2, CH_4.

Sample Questions

- Which molecules did you pick as the ones that didn't have a smell? What was your reasoning?
- Why doesn't carbon dioxide have a smell? What about methane?
- What does the overall symmetry of a molecule have to do with its smell?
- Can some molecules be symmetrical in shape and still be polar? Explain.

Key Points

Small nonpolar molecules do not have a smell. The most obvious examples of small nonpolar molecules are molecules with two identical atoms, such as oxygen, O_2, and nitrogen, N_2. Molecules with more than two atoms can also be nonpolar if they have a symmetrical shape. For example, tetrafluoromethane, CF_4, is nonpolar. Although all the fluorine atoms are highly electronegative, the entire molecule is a perfectly symmetrical tetrahedron. No portion of the molecule is any more electronegative than any other portion.

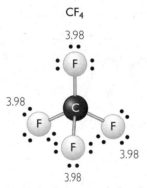

Tetrafluoromethane is symmetrical and nonpolar.

If the overall shape of a molecule is asymmetrical and the molecule is made from more than one kind of atom, chances are it is a polar molecule. If the overall shape of a molecule is highly symmetrical, it is necessary to look closer to see whether any imbalance in electronegativity is still present. The presence of different atoms on different sides of the molecule, as in chlorotrifluoromethane, is a sign of polarity.

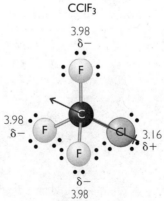

Chlorotrifluoromethane is polar because of the chlorine atom on one side.

Discuss What Might Be Happening in the Nose

➡ Ask students to show the class how a molecule of ammonia might interact with molecules of water.

→ You might want to have a model of a molecule such as citronellol available to discuss the polarity of molecules of this size.

Sample Questions

- Why do you think polarity is important to smelling? What does it have to do with the nose? With receptor sites?
- What is the inside of the nose like? Do you think the inside of the nose is polar? Explain.
- Do you think a larger molecule like citronellol is considered a polar molecule? Does it dissolve in water? Explain.

Key Points

Inside the nose is a watery mucous lining. One theory is that polar molecules dissolve in the mucous lining and then attach to receptor sites in the nose. Because of intermolecular attractions, polar molecules dissolve very easily in other polar molecules. Nonpolar molecules, on the other hand, do not dissolve in water or in other polar molecules. This might explain why we cannot smell small nonpolar molecules: They cannot move through the mucous lining to the receptor sites. Ammonia dissolves in water.

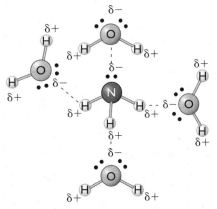

The intermolecular attractions of polar molecules cause them to dissolve easily in water.

Molecules need to be attracted to receptor sites in order to be detected. Another smell theory is that polar molecules "dock" in polar receptor sites after they enter the nose. Receptor site molecules have polar aspects to them, allowing a smell molecule to "attach" to the site. Nonpolar molecules, such as methane and carbon dioxide, would not dock in receptor sites. Medium-size molecules that smell, like citronellol, usually are asymmetrical in shape and often have lone pairs somewhere on the molecule. They are polar enough to dock in receptor sites.

The small molecules that constitute our air do not have a smell. They can all be shown to be nonpolar. Presumably, they simply move into our nostrils and breathing passages and then move out again. One exception is water, which is a small polar molecule that does not have a smell. Scientists believe that this is because there is so much water present at all times in our nose that the signal to our brain never changes, and hence we do not smell it. It is as if we have become so used to it that it is like background noise. Many animals, including elephants, seem to be able to smell water at great distances.

Wrap-up

Key Question: What does polarity have to do with smell?

- Differences in electronegativity values can be used to determine the direction of polarity of an entire molecule. (In other words, you can determine which part of the molecule has a partial negative charge and which part has a positive partial charge.)
- Molecules that are asymmetrical in shape and composition are usually polar and usually smell.
- Small polar molecules have a smell. Small nonpolar molecules do not have a smell.
- Polar molecules dissolve easily in other polar molecules. Nonpolar molecules do not dissolve easily in polar molecules.

Evaluate (5 minutes)

> **Check-in**
>
> Is hydrogen cyanide, HCN, a polar molecule? Will it smell? Why or why not?
>
>

Answer: Even though HCN is a linear molecule, you can tell at a glance that it is an asymmetrical molecule. It has a nitrogen atom on one end and a hydrogen atom on the other end. The nitrogen atom is more electronegative than the hydrogen atom. Thus, the nitrogen end has a partial negative charge on it. This molecule smells—a little like almonds.

Homework

Assign the reading and exercises for Smells Lesson 18 in the student text.

LESSON 19 OVERVIEW

Sniffing It Out
Phase, Size, Polarity, and Smell

Lesson Type
Classwork: Pairs

Key Ideas

Broad generalizations can be made relating smell to polarity, molecular size, phase, and type of bonding. For instance, most medium-size molecules—those with about 5 to 19 carbon atoms—have a smell, including nonpolar ones; small molecules tend to have a smell only if they are polar; and large molecules generally do not have a smell. Ionic and metallic solids do not vaporize and therefore cannot be smelled.

As a result of this lesson, students will be able to

- explain the connections between smell and polarity, molecular size, phase, and type of bonding
- predict whether a molecule will have a smell based on its structure, composition, and phase

Focus on Understanding

- Some substances seem to have a smell, but we really smell contaminants. For example, students might claim that metals have a smell. The smells coming from metal objects usually are oils or other molecular substances. Plastics seem to smell, but we are really smelling compounds used in their manufacture.
- Carbon atoms can form four bonds allowing for large molecules. The number of carbon atoms is often used to indicate molecular size.

What Takes Place

Students are given a worksheet with information about how polarity, size, phase, and type of bonding are related to whether a substance will have a smell. Students refine their hypotheses about smell and put smell into a broader context.

Materials

- student worksheet
- transparencies—Three Phases and Check-In
- Smells Summary Chart from Lesson 13: Sorting It Out
- large sheet of paper for making a No Smell Summary Chart

Setup

Post the Smells Summary Chart and blank paper for the No Smell Summary Chart.

LESSON 19 GUIDE

Sniffing It Out
Phase, Size, Polarity, and Smell

Engage (5 minutes)

Key Question: What generalizations can you make about smell and molecules?

ChemCatalyst

1. If you place an open perfume bottle and a piece of paper in a sunny window, the aroma of the perfume will soon fill the air, but you won't smell the paper at all. Explain what is going on.
2. What is the heat from the Sun doing to the perfume to increase the smell?

Sample Answers: 1. Students should relate heating to phase change and will offer explanations based on the movement of molecules and molecules becoming airborne. Students might say that in order to smell a substance, the substance must become a gas. 2. The heat from the Sun provides enough energy to overcome the intermolecular forces that are keeping the perfume in liquid form.

Discuss the ChemCatalyst

Sample Questions

- Why can you smell a solid substance, like a cookie, when it's being cooked in the oven?
- If a plastic bag, a piece of paper, or a book warms up in the Sun, will you suddenly be able to smell it? Why or why not?

Explore (15 minutes)

Introduce the Classwork

➡ Students can work in pairs. Point out that the range for molecular size is small—fewer than 5 carbon atoms; medium—5 to 19 carbon atoms; large—20 or more carbon atoms.

LESSON 19 CLASSWORK

Sniffing It Out
Phase, Size, Polarity, and Smell

Name _____
Date _____ Period _____

Purpose
To analyze data that relate the size of molecules, type of bonding, phase, and polarity to smell.

Instructions
In this data table, substances that have no smell are shaded gray. Use the data table to answer the questions. In general, molecules with around 5 to 19 carbon atoms are considered medium-size.

Substance type and bonding	Size	Smell?	Phase	Examples Name	Formula
Molecular Nonpolar covalent	small molecules	no	gas	nitrogen oxygen carbon dioxide methane	N_2 O_2 CO_2 CH_4
Molecular Polar covalent	small molecules	yes	gas	hydrogen chloride hydrogen sulfide ammonia fluoromethane	HCl H_2S NH_3 CH_3F
Molecular Polar and nonpolar covalent	medium-size molecules	yes	liquid	octane geraniol carvone pentyl propionate	C_8H_{18} $C_{10}H_{18}O$ $C_{10}H_{14}O$ $C_8H_{16}O_2$
Molecular covalent	large molecules	no	solid	1-triacontyl palmitate (beeswax) polystyrene cellulose	$C_{46}H_{92}O_2$ $C_{8000}H_{8000}$ $C_{1800}H_{3000}O_{1500}$
Ionic metals bonded to nonmetals	N/A	no	solid	sodium chloride (table salt) calcium oxide (lime) calcium carbonate (chalk)	$NaCl$ CaO $CaCO_3$
Metallic only metal atoms	N/A	no	solid	gold copper aluminum brass	Au Cu Al $CuZn$

Living By Chemistry Teaching and Classroom Masters: Units 1–3
© 2010 Key Curriculum Press

Unit 2 Smells **183**
Lesson 19 • Worksheet

Questions

1. Which types of gaseous molecules have a smell? *polar covalent molecules*
2. What types of molecules made from nonmetals have a smell?

   small polar covalent, medium-size polar and nonpolar covalent molecules
3. Do nonpolar molecules have a smell?

   Not small ones, but medium-size ones do.
4. If something is a solid, do you expect it to have a smell? How do you explain this?

   No. It must become a gas to get into the nose.
5. If a substance contains a metal, do you expect it to have a smell? Explain.

   No. It will have either ionic or metallic bonding. Those substances don't have a smell.
6. Describe the types of substances that were in the vials in this unit.

   liquids; medium-size molecular covalent molecules; polar and nonpolar molecules
7. Put an X in column 1 or 2, then complete the table.

Will smell	Won't smell	Example	Chemistry reasoning
	X	a brass doorknob—Cu, Zn	Metallic substance.
X		sweaty socks—hexanoic acid, $C_6H_{12}O_2$	Medium-size, covalent molecule. Is a carboxylic acid.
	X	epsom salts—magnesium sulfate, $MgSO_4$	Ionic solid. Doesn't become a gas.
X		anisyl alcohol in laundry soap—$C_8H_{10}O_2$	Medium-size, covalent molecule. An alcohol.
	X	plastics, DNA, proteins, starch, paraffin, cellulose	Solid substances made of large molecules that will not easily become gases.
	X	sunflower oil, $C_{21}H_{38}O_5$	Too large a molecule to become airborne.

8. **Making Sense** In general, what kinds of substances have a smell?

   Answers will vary. See the list in the Explain and Elaborate section.

Explain and Elaborate (20 minutes)

Come Up with a Comprehensive Model for Smell

➥ There are many good ways to summarize this information. You could write the headings "polarity," "phase," and "molecular size" on the board and write student generalizations beneath the appropriate headings.

Sample Questions

- When determining smell, when is it important to consider the polarity of a molecule? (when small molecules are being considered)
- What is a medium-size molecule according to the data table? (a molecule with around eight to ten carbon atoms)
- What did you discover about phase and molecular size?
- Now that you've seen all of this information, what determines whether a substance smells?

Key Points

Phase and molecular size both play a role in smell properties. Medium-size molecules all seem to have a smell, and they are liquids or gases. Their smell is determined by shape and functional group. Large molecules do not have a smell. They are too big and bulky to become gases and move into the nose.

Polarity determines the smell of small molecules. Small polar molecules have a smell. Small nonpolar molecules do not have a smell.

Many solids do not evaporate into gases and therefore don't have a smell. Solids tend not to smell unless they can become gaseous (e.g., components of a chocolate bar). Nonmolecular solids (ionic and metallic solids) do not have a smell.

Relate Phase, Bonding, and Smell

[T] ➥ Display the transparency Three Phases showing methane, octane, and polystyrene to assist you in guiding the discussion.

Sample Questions

- What does phase have to do with the process of smelling? (The substance needs to be in the gas phase.)
- How is the type of bonding related to phase? (Mainly molecular substances vaporize at room temperature.)
- How is the size of a molecule related to its phase? (As molecules get bigger, they tend to be solids at room temperature.)
- What are the attributes of molecules that have a smell? (Less than about 20 carbon atoms, mainly molecules with carbon; small molecules need to be polar.)
- Why has our study of smells focused on molecules? (Ionic, network covalent, and metallic solids generally are not gases at 25 °C, and thus they have no smell because they cannot enter the nose.)

Key Points

Molecular substances tend to have a smell because it is easy for them to become airborne. A substance has to be airborne to get into the nostrils and be detected by the nose. If a substance is a gas at 25 °C, then it probably is composed of molecules. Methane is a good example of a molecule that is a gas at 25 °C. However, methane does not have a smell because it is not a polar molecule. Note that a substance that smells, such as dimethyl sulfide or butane thiol, is usually added to methane to help in detecting leaks.

Methane, $CH_4(g)$

Substances that are liquids at ordinary temperatures tend to have a smell. Gasoline is an example of a medium-size molecule that is a liquid and vaporizes easily, and therefore can be smelled. Even though gasoline is nonpolar, it does have a smell. Most molecules with 5 to 20 carbon atoms are liquids at 25 °C and have a smell.

Gasoline, $C_8H_{18}(l)$

Molecular solids are volatile and have a smell. You know from experience that some solids sublime; that is, a few molecules go directly from the solid phase to the gas phase without passing through the liquid phase. These substances consist of molecules, and they do have a smell. This is how you are able to smell a piece of chocolate or many of the foods you prepare. In contrast, ionic, network, and metallic substances do not smell. They do not consist of molecules, so there are no small units that can enter the gas phase readily. Polystyrene is an example of a solid that does not have a smell. You can think of it as consisting of long chain molecules or as a network covalent solid.

Polystyrene, $(C_8H_8)_n(s)$

Update the Smells Summary Chart

▶ Add the new information to the existing Smells Summary chart. Use a second piece of paper to create a new chart that says "no smell" in the middle.

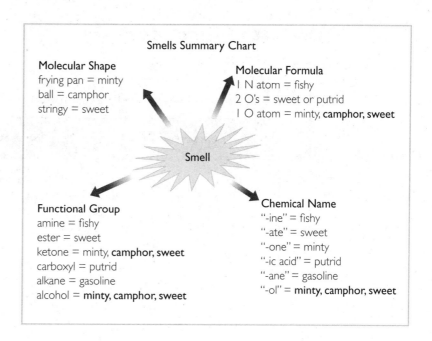

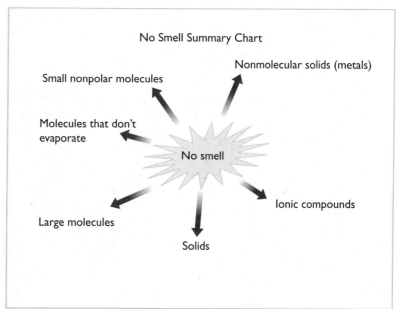

Wrap-up

Key Question: What generalizations can you make about smell and molecules?

- Small molecules have a smell if they are polar.
- Medium-size molecules tend to have a smell whether or not they are polar.
- The smells of medium-size molecules can be predicted by looking at shape and functional group.
- Very large molecules do not have a smell because they do not evaporate and enter the nose.
- Nonmolecular solids (e.g., salts, metals) do not have a smell because they do not evaporate.

Evaluate (5 minutes)

Check-in

Which of these will have a smell? Explain your reasoning.

Substance	Structure	Phase
$CaCl_2$, calcium chloride	Cl^- Ca^{2+} Cl^- (repeating throughout the solid in three dimensions)	solid
$C_8H_8O_3$ vanillin	(structural diagram of vanillin)	liquid
HCN, hydrogen cyanide	H—C≡N	gas

Answer: Calcium chloride will not have a smell. It is an ionic compound and is also a solid. Vanillin probably will have a smell. It is a medium-size molecule (in fact, it smells like vanilla). Hydrogen cyanide also has a smell. It is a small polar molecule.

Homework

Assign the reading and exercises for Smells Lesson 19 and the Section III Summary in the student text.

SECTION IV

Molecules in the Body

The final section of the Smells unit focuses on biological molecules. Lesson 20 covers the "handedness" of molecules and challenges students to visually and structurally distinguish between mirror-image isomers. Lesson 21 introduces amino acids and shows students how to connect and fold them to form a protein. This lesson also revisits the receptor site, or "lock-and-key," mechanism that characterizes the sense of smell. Lesson 22 summarizes the context of smell and all that has been learned about molecules.

In this section, students will learn

- how to identify mirror-image isomers
- how "handedness" affects molecular properties
- about the formation of proteins from amino acids
- about basic "lock-and-key" mechanisms in biological processes

LESSON 20 OVERVIEW

Mirror, Mirror
Mirror-Image Isomers

Lesson Type
Activity:
Groups of 4

Key Ideas

Mirror-image isomers are a pair of molecules that differ only in that they are the mirror image of each other. These molecules are said to have a "handedness" because they differ from one another the way a right hand and left hand differ. Mirror-image isomers can have different properties from each other. For example, they can have different smells.

As a result of this lesson, students will be able to

- recognize mirror-image molecular structures
- explain what it means for molecules to be superimposable
- understand why mirror-image isomers have different properties

Focus on Understanding

- It is sometimes difficult to visualize how mirror-image isomers differ from each other. Demonstrate with models as much as possible.

- Students might need assistance in understanding what it means that two molecules are superimposable. Show them that your hands are not superimposable. In other words, no matter how you turn and twist your left hand, it will never have the same orientation as your right hand.

- Be aware that some individuals cannot detect the difference in smell between D-carvone and L-carvone. This "teaching moment" is an opportunity to point out that there often are exceptions to the rule. These exceptions do not necessarily invalidate the hypothesis, but they do indicate that there is more to learn about smell receptors.

Key Term

mirror-image isomer

What Takes Place

Students make models of several tetrahedral structures and view their images in a mirror. Students then build the mirror-image model of each molecule and explore whether the pair of molecules are superimposable on each other. They compare the smell of L-carvone (which they smelled in Lesson 1) with that of D-carvone and explain the nature of the mirror images of these molecules.

Materials

- student worksheet
- D-carvone (caraway seed oil), 5 mL
- pair of gloves, mittens, or shoes (optional)
- space-filling models of L-carvone and D-carvone
- computer and projection system for showing interactive computer-simulated images (such as Rasmol or Chime or Jmol) of L-carvone and D-carvone (optional)
- larger mirror (about 8 in. by 10 in.) for demonstrating molecules at the front of the room

Per groups of 4

- molecular modeling set or plastic bag containing 4 black spheres (4-hole), 8 white spheres, 4 red spheres, 2 blue spheres, and 7 straight connectors.
- hand mirror (3 in. by 3 in.)
- vials A and Z (L-carvone and D-carvone)
- small piece of masking tape

Setup

Vial A from Lesson 1 contains L-carvone. Prepare vial Z by first placing a cotton ball in the vial. Use a plastic pipette to deliver three to five drops of the caraway seed oil. (Ground-up or crushed caraway seeds can be substituted if they have a strong enough smell.) Place sets of the two vials in plastic sandwich bags for ease of distribution.

Build one or more sets of *space-filling models* of L-carvone and D-carvone with the molecular model sets provided in the kit for students to examine as they are working on the worksheet. Use the disc connectors to represent bonds (double discs for double bonds) and white hemispheres for hydrogen. Put a tag on each with the correct name. Ball-and-stick models and structural formulas are shown here for reference. You may want to build one isomer and use a mirror to construct the other.

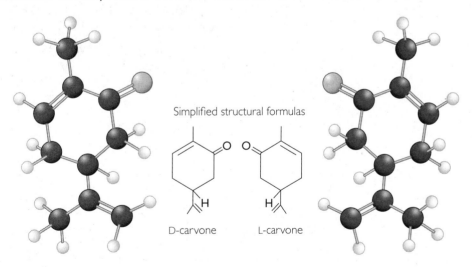

Cleanup

Save the vials for reuse in other classes. When you are finished using all the vials in all of your classes, remove the cotton balls from the vials, place them in a plastic bag, and dispose of them. Let the vials air out in a hood or rinse them with acetone for reuse the next time you do this unit. If you do not remove the cotton balls, the substances will mix with each other and begin to smell putrid.

LESSON 20 GUIDE

Mirror, Mirror
Mirror-Image Isomers

Engage (5 minutes)

Key Question: What are mirror-image isomers?

> **ChemCatalyst**
>
> Which of these objects look identical in a mirror? Which look different? Explain any differences.
>
> 1. glove   2. barbell   3. spring   4. tetrahedron

Sample Answers: 1. A glove does not look identical in a mirror. When both thumbs point in the same direction, the palms are opposite. 2. A barbell looks identical in a mirror. Both "sides" are the same (but any writing will be reversed). 3. A spring does not look identical in a mirror. The mirror image of a spring has the opposite rotation. If one twists clockwise, the other twists counterclockwise. 4. Some students might be uncertain about a tetrahedron. Accept their guesses. They will have an opportunity to test their ideas in the activity.

Discuss the ChemCatalyst

➡ Assist students in describing mirror images.

Sample Questions

- What other objects look different from their mirror image? (feet, ears, shoes, car doors, side-view mirrors, baseball mitts, golf clubs)
- What objects don't look different from their mirror images? (symmetrical objects: cowboy hats, rearview mirrors, etc.)
- What does it mean that something has a "handedness"?
- If a molecule and its mirror image are different, do you think the two molecules will have the same smell? Why or why not?

Explore (20 minutes)

Introduce the Activity

➡ Introduce the concept of handedness through a few quick demonstrations.

- Hide both hands behind a binder or a large piece of paper. Extend one hand. Ask students to identify it as left or right. Ask them how they could tell.
- Ask students what a receptor site might look like for each hand and how the sites might be different. (A glove would be a receptor site metaphor.)

➡ Pass out the student worksheets, mirrors, and molecular model kits.

Guide the Activity

➡ Suggest that students view the mirror from behind the model (not from the side). It will be more obvious whether the images are identical.

➡ Be sure students clearly understand that superimposable means you can place two objects on top of each other so that they look identical in every way. You cannot superimpose your left hand on your right hand. Either one hand will be palm up and the other palm down, or your thumbs will be pointing in opposite directions.

LESSON 20 ACTIVITY

Mirror, Mirror
Mirror-Image Isomers

Name _____
Date _____ Period _____

Purpose
To understand how mirror-image molecules can have different properties.

Materials
- molecular model kit
- vials A and Z
- small piece of masking tape

Instructions

1. Build a model of CH_4. Use a black sphere for the carbon atom and white spheres for the hydrogen atoms.
   a. Compare this model with its image in the mirror. Write the similarities and differences in the table.
   b. Build a second molecule that looks like the mirror image. Determine whether the mirror image can be superimposed on the original image. Enter your answer in the table. Take apart the models.
2. Repeat for CH_3F. Use a red sphere for the fluorine atom.
3. Repeat for CHFClBr. Use a red for the fluorine, a blue for the chlorine and a red with a piece of tape on it for the bromine atoms.
4. Repeat for $C(CH_3)HFCl$. Attach the CH_3 (called a methyl group) to the central carbon. Use a red for the fluorine and a blue for the chlorine.

Molecule	Compare with mirror image		Can the second molecule be superimposed on the first?
	What is the same?	What is different?	
CH_4	same atoms, connected the same	nothing	yes
CH_3F	same atoms, connected the same	nothing	yes
CHFClBr	same atoms, connected the same	different order	no
$C(CH_3)HFCl$	same atoms, connected the same	different order	no

Analysis

1. When a molecule and its mirror-image cannot be superimposed on each other, they are called mirror-image isomers.
   a. Which molecules in the table have mirror-image isomers?

   CHFClBr and $C(CH_3)HFCl$

b. What do these two molecules have in common?

Four different atoms (or group of atoms) are connected to the carbon atom in the center.

2. Carvone has the molecular formula $C_{10}H_{14}O$. The two mirror-image isomers are shown here as ball-and-stick models next to simplified structural formulas.

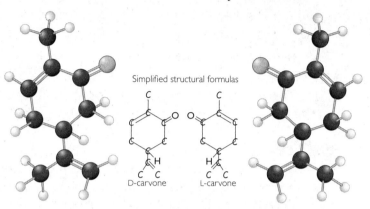

a. In the simplified structural formulas, label the location of each carbon atom with a C. What is missing from the simplified structural formulas?

Most of the hydrogen atoms are not shown.

b. Examine the molecular models provided by your teacher. Are the mirror images superimposable on each other? Explain why or why not.

No. If the O atoms are superimposed, the H atom near the middle of the molecule points toward you instead of away from you. (The C(CH_2)(CH_3) groups at the bottom also point in opposite directions.)

c. The L-carvone molecule smells minty. Do you expect D-carvone to smell minty also? Explain your thinking.

Sample answers: Both are ketones, so they should smell minty. But they are constructed differently, so they may smell different.

d. Smell the contents of vials A and Z. Record your observations.

They smell different. Vial A smells like spearmint. Vial Z smells like dill or caraway seeds (rye bread).

3. Making Sense Explain why mirror-image isomers have different smells.

Mirror-image isomers must fit into different receptor sites, just as your left foot does not fit into a right shoe.

4. If You Finish Early Does difluoromethane, CH_2F_2, have a mirror-image isomer? Explain your thinking. Build a model to check your answer.

CH_2F_2 does not have a mirror-image isomer. No matter how you construct it, it is superimposable on its mirror image.

Explain and Elaborate (15 minutes)

Discuss Mirror-Image Molecules

→ Have space-filling models of the molecules from the worksheet available to demonstrate the concepts. A large mirror is also helpful.

Sample Questions

- How can you determine whether a molecule has a mirror-image isomer?
- How does the mirror help you with the activity? Explain what it tells you.
- What did you discover when you made models of the mirror images?
- Look at the two carvone molecules on the worksheet. In each, the carbon atom at the bottom of the carbon ring is attached to four different "things." What happens to these four things in the mirror image?

Key Points

Mirror-image molecules that cannot be superimposed on each other are called mirror-image isomers.   Your left hand and your right hand are mirror images of each other. They cannot be superimposed on each other. Similarly, CHFClBr and C(CH$_3$)HFCl have mirror-image isomers. The mirror images are not superimposable; thus, they are distinct molecules. Molecules also can have mirror images that *can* be superimposed on each other. These are not considered mirror-image isomers because they are both identical. Methane, CH$_4$, and fluoromethane, CH$_3$F, are examples.

Tetrahedral molecules in which four *different* atoms or groups are attached to a carbon atom always have mirror-image isomers.   Indeed, any molecule that contains even one carbon atom with four *different* atoms or groups attached to it will have a mirror-image isomer. The carbon atom at the bottom of the carbon ring in the carvone molecule on the worksheet is an example.

Discuss Properties of Mirror-Image Isomers

Sample Questions

- What did you conclude about D-carvone and L-carvone?
- How are D-carvone and L-carvone similar? How are they different?
- Do you expect D-carvone and L-carvone to fit differently into receptor sites? Why or why not?

Key Points

The mirror-image isomers D-carvone and L-carvone have different smells. Thus, they probably do not fit into the same receptor site. Just as a left hand does not fit into a right-hand glove, mirror-image isomers do not fit into the same lock-and-key sites. This means that they can have different properties, including smell.

The mirror-image isomers have a "handedness."   Human hands are symmetrical but different from each other. No matter how hard you try, you cannot superimpose a right hand onto a left hand so that the two are identical. The thumbs will always be on opposite sides, or the palms will be on opposite sides. Mirror-image isomers are molecules that have a "handedness." Just as your right hand and left hand need different gloves, mirror-image isomers may need different receptor sites.

Wrap-up (5 minutes)

Key Question: What are mirror-image isomers?

- Molecules that are not identical to their mirror image are called mirror-image isomers.
- Mirror-image isomers have a "handedness," just like a right hand and a left hand.
- Mirror-image isomers can have different properties, including smell.
- Molecules that have at least one carbon atom with four different atoms or groups attached to it have a mirror-image isomer.

Evaluate (5 minutes)

Check-in

Which of these molecules will have a mirror-image isomer? Explain your reasoning.

A. CF_4 **B.** CHF_3 **C.** $C(CH_3)_4$

Answer: None of these molecules will have a mirror-image isomer. These molecules are superimposable on their mirror images, and the mirror images represent the same molecule. They do not have a carbon atom with four different things attached to it.

Homework

Assign the reading and exercises for Smells Lesson 20 in the student text.

LESSON 21
Protein Origami
Amino Acids and Proteins

OVERVIEW

Key Ideas

Lesson Type
Activity:
Groups of 4

Smell receptor sites are composed of proteins, which are made from a chain of amino acids. One end of an amino acid is an amine functional group, and the other end is a carboxyl group. The amine group from one amino acid links with the carboxyl group of another amino acid to form proteins. There are 20 amino acids used by the body. Of the two mirror-image isomers of each amino acid, only the left-handed one is usable in the human body.

As a result of this lesson, students will be able to

- explain that protein molecules are chains of amino acid molecules
- understand that the smell receptor sites are protein chains folded to form a receptor of a specific shape
- explain the "handedness" of a smell receptor site

Focus on Understanding

Topic: Proteins
Visit: www.SciLinks.org
Web code: KEY-221

- It is useful to point out that receptor sites are not mirror images of molecules but rather a "print" of the molecule itself. For example, your right foot is not a "receptor site" for your left foot. A left footprint is.
- A protein is a chain of over 100 amino acids. A shorter chain typically is referred to as a polypeptide. However, in the interest of simplicity, we do not make a clear distinction in this lesson.

Key Terms

amino acid   protein   peptide bond (amide bond)

What Takes Place

Students examine the structures of amino acid molecules. They build the mirror-image isomer of the amino acid molecule alanine used by the body. Students then build two other amino acid molecules and link them to form a dipeptide. In the Explain and Elaborate discussion, the class examines a model of a protein with six amino acid molecules linked. The protein is folded to form a smell receptor site.

Materials

- student worksheet
- handout—Twenty Amino Acids
- transparencies—ChemCatalyst, Amino Acids
- space-filling models of L-carvone and D-carvone (from previous lesson)
- moist sand or soft clay, enough to make an impression of the space-filling models of D-carvone and L-carvone (optional)

- tray to hold sand or clay (optional)
- ball-and-stick models of any six amino acid molecules (see Twenty Amino Acids handout)

Per groups of 4

- molecular modeling set or plastic bag containing 6 black spheres (4-hole), 4 red spheres, 2 blue spheres, 8 white spheres, 11 straight connectors, and 4 curved connectors.

Setup

If you have enough modeling sets, you might want to build a few ball-and-stick models from the handout ahead of time. Some of these can be the same. All should be the "left-handed" mirror-image isomer. You can tell that you have the left-handed isomer by viewing the molecule with the H atom pointing toward you. The other three groups should be positioned with COOH, the R group, and NH_2 in clockwise order. You can remember this with the acronym CORN, where CO stands for the COOH group, R for the R group, and N for the NH_2 group.

Link these amino acids to form a protein molecule.

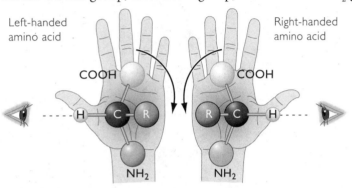

You will be using this model after the activity to demonstrate how a protein molecule folds to form a receptor site. You can practice ahead of time twisting it so it coils like a helix and then into a fairly compact bundle with an indentation. If you don't have enough modeling materials for each group to build two amino acid molecules, you can join pairs or groups to link their amino acids together for Question 3 on the worksheet. You can do the If You Finish Early at the front of the room, with groups linking their amino acids together to make a protein and fold it into a receptor site.

Cleanup

Allow time for students to take apart the protein molecule in preparation for the next class.

LESSON 21
Protein Origami
Amino Acids and Proteins

GUIDE

Engage (5 minutes)

Key Question: What is a receptor site made of?

ChemCatalyst

The mirror-image isomers of carvone are shown.

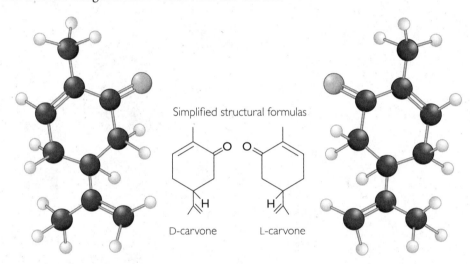

1. Explain how the receptor sites for D-carvone and L-carvone might be different from each other.
2. Sketch receptor sites for D-carvone and L-carvone. Assume that the polar side (the side with the oxygen atom) attaches to the receptor site.

Sample Answers: 1. Because the molecules are mirror images of each other, the receptor sites for these molecules probably are mirror images of each other as well. 2. The receptor site probably is relatively narrow, because the molecule enters with the flatter part perpendicular to the receptor site. For D-carvone, the bottom of the receptor site bends to the right to accommodate the C(CH$_2$)(CH$_3$) group. For L-carvone, the bottom bends to the left.

Discuss the ChemCatalyst

- Ask one or two students to draw receptor sites for D-carvone and L-carvone on the board.
- Optional: Use a tray with sand or soft clay to make "molecule prints" of D-carvone and L-carvone. Show that each isomer fits only into its own molecule print, just as your right foot fits only into its right footprint.

Sample Questions

- What would receptor sites for your left hand and right hand look like?
- How is a handprint of your left hand different from one of your right hand?
- What would you observe if you made "molecule prints" of D-carvone and L-carvone?
- Some people cannot smell L-carvone at all. What might explain this?

Explore (20 minutes)

Introduce the Activity

- Give students the handout Twenty Amino Acids.
- Explain that amino acids are molecules the body uses to make proteins. Proteins are long chains of amino acids. There are 20 amino acid molecules that the human body uses. Each has a mirror-image isomer. The human body can use only the left-handed mirror-image isomer.

> **Amino acid:** A molecule with a carboxyl functional group and an amine functional group.
>
> **Protein:** A large molecule consisting of amino acids bonded together.

- Explain R groups. Aside from the carboxyl functional group, the amine functional group, and the hydrogen atom, an amino acid has a fourth group that identifies it, called the R group. On the handout, the R groups are shown at the bottom of each molecule (everything below the second carbon atom).
- Explain how to tell that an amino acid is left-handed. You can tell that you have the left-handed isomer by viewing the molecule with the H atom pointing toward you. The other three groups should be positioned with COOH, the R group, and NH_2 in clockwise order. You can remember this with the acronym CORN, where CO stands for the COOH group, R for the R group, and N for the NH_2 group.

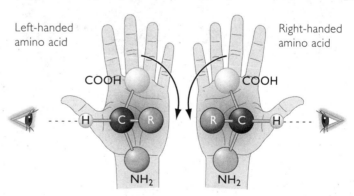

- Tell students that they will be working with amino acids to build protein molecules that fold into smell receptor sites.

- Pass out the student worksheets and molecular model kits to groups of four students.

LESSON 21 ACTIVITY

Protein Origami
Amino Acids and Proteins

Name _____
Date _____ Period _____

Purpose
To understand how amino acid molecules link to form proteins.

Materials
- molecular model kit

Instructions and Questions

1. Examine the amino acid molecules on the handout.

   a. What do all the amino acids have in common? How do they differ?

   They all have a carbon atom that is attached to a COOH group, to an NH₂ group, and to an H atom. The fourth group attached to this carbon atom differs for each amino acid.

   b. Some of the amino acids are labeled as hydrophilic, or "water-loving." Explain why these amino acids are attracted to water.

   For the hydrophilic amino acids, the R groups have a nitrogen or an oxygen atom. Thus, the R groups are polar.

   c. Some of the amino acids are classified as hydrophobic, or "water-fearing." Explain why these amino acids are not attracted to water.

   For the hydrophobic amino acids, the R groups consist mainly of carbon and hydrogen. Thus, the R groups are nonpolar.

2. Use the structural formula and the model kit to build a ball-and-stick model of the amino acid alanine. Arrange the alanine molecule so that you have the left-handed isomer (switch the groups if necessary).

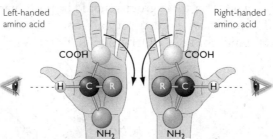

Left-handed amino acid

Right-handed amino acid

If you look at the molecule with the H atom pointing toward you, the COOH group, the R group, and the NH₂ group should be arranged clockwise.

Sketch your model here. Label the oxygen atoms O and the nitrogen atom N.

Living By Chemistry Teaching and Classroom Masters: Units 1–3
© 2010 Key Curriculum Press

Unit 2 Smells   **189**
Lesson 21 • Worksheet

3. Amino acid molecules can link together in long chains to form large protein molecules. The bond between two amino acids is called an amide bond or a peptide bond.

   a. Use the model kit to build the left-handed isomer of two amino acids on the handout (other than alanine).

   b. Assemble a molecule from the two amino acids, as shown here.

   [Diagram showing two amino acids combining with a peptide bond, releasing H_2O]

   c. Draw the structural formula of three amino acid molecules linked together.

   [Diagram showing three amino acids linked with R_1, R_2, R_3 side chains]

   d. How many water molecules are produced when you link three amino acid molecules together?

   two, one for each of the two linkages

   e. Does the HONC 1234 rule apply to these larger molecules? *yes*

4. Most protein molecules do not remain stretched out. Instead, they tend to fold into globular shapes, because an amino acid molecule in one part of the chain can attract an amino acid molecule in another part of the chain.

   a. How might protein folding explain how proteins create smell receptor sites?

   Protein molecules can twist and turn so that a pocket is created to receive a smell molecule.

   b. How might protein molecules account for the great variety of smell receptor sites?

   Different receptor sites probably consist of proteins with different sequences of amino acids.

5. **Making Sense** How can ideas about amino acids and proteins help you understand how people detect smells?

   The proteins fold to form specific smell receptor sites. Maybe molecules that we smell are attracted to these receptor sites.

6. **If You Finish Early** Build a protein molecule with three or four amino acid molecules. Examine how the molecule might fold to form a receptor site.

TWENTY AMINO ACIDS

Leucine   Isoleucine   Valine

Asparagine   Methionine   Phenylalanine

Glutamic Acid   Glutamine   Histidine   Aspartic Acid

Serine   Threonine   Lysine

Arginine   Tyrosine   Tryptophan

Cysteine   Alanine   Glysine   Proline

Explain and Elaborate (15 minutes)

Discuss Amino Acids

Sample Questions

- What are the main features of amino acid molecules?
- Explain why amino acid molecules all have mirror-image isomers.
- How do amino acid molecules link together?
- Show that the HONC 1234 rule applies to a dipeptide.

Key Points

Amino acids are molecules with a carboxyl group, COOH, and an amine group, NH_2. There are hundreds of naturally occurring amino acids. Each amino acid has a mirror-image isomer because four different groups are attached to a central carbon atom. Your body uses 20 different amino acids, and it uses the left-handed isomers exclusively.

Amino acids can link together to form long chains called proteins. The carboxyl group of one amino acid can link to the amine group of another amino acid with a peptide bond. Water is lost in the process. In this way, it is possible to form long chains of amino acids called proteins. It is possible to make a wide variety of proteins by linking amino acids in different sequences. Your hair, your skin, enzymes, and smell receptor sites are all made of proteins with different sequences of amino acids.

Discuss the Folding of Proteins

➡ Use your ball-and-stick models of about six amino acid molecules linked together. Demonstrate how this chain of amino acids can be twisted to form a smell receptor site.

Sample Questions

- Why do you think protein molecules tend to fold into globs rather than remain stretched out?
- Why are amino acids in different parts of the protein attracted to one another?
- Which amino acids are likely to be attracted to polar R groups?
- How can protein molecules account for the variety of smell receptor sites in the nose?

Key Point

Protein molecules fold to form pockets because of attractions between the R groups on different amino acids in the chain. The polar R groups tend to attract other polar R groups. These hydrophilic (water-loving) R groups tend to be on the outside of the molecule, near the watery environment. The hydrophobic (water-hating), nonpolar R groups tend to cluster in the center of the molecule. The lining of the nose is made of protein molecules. Each kind of protein folds in a different way to create a smell receptor site for a different smell molecule. When a smell molecule attaches to the site, a nerve carries a signal from that site to the brain.

Wrap-up

Key Question: What is a receptor site made of?

- Amino acids are molecules with a carboxyl group, COOH, and an amine group, NH_2.
- Amino acids all have mirror-image isomers. Only left-handed amino acid molecules function in the human body.
- The carboxyl group from one amino acid can link with the amine group of another amino acid forming a peptide bond.
- Proteins are long chains of amino acids linked together.
- Protein molecules can fold to form smell receptor sites.

Evaluate (5 minutes)

> **Check-in**
>
> Name two concepts from the Smells Unit that were used today to help you understand proteins.

Answer: Possible answers: the HONC 1234 rule, structural formulas, polarity, attractions between polar molecules, and so on.

Homework

Assign the reading and exercises for Smells Lesson 21 and the Section IV Summary in the student text.

LESSON 22 OVERVIEW

Who Nose?
Unit Review

Lesson Type
Classwork: Pairs

Key Ideas

Students review what they have learned in Unit 2: Smells regarding molecular structure and its relationship to the property of smell. Molecular formula, structural formula, functional group, three-dimensional shape, polarity, molecular size, molecular phase, and handedness all provide relevant information that can be taken into account in determining the smell of a molecule.

As a result of this lesson, students will be able to

- ask clarifying questions regarding concepts covered in this unit
- create a list of topics and concepts to study for an upcoming exam

Focus on Understanding

- It is important for students to understand that there probably are exceptions to all our generalizations about smells. Nevertheless, the broad generalizations still hold true and still are valuable.

What Takes Place

Using a worksheet, students review the unit. Students apply and integrate what they've learned about smells and chemistry. They draw structural formulas from molecular formulas, identify functional groups, and predict the smells and properties of compounds from assorted chemical information. Finally, they predict the smell of a compound and then test the accuracy of their prediction. You might want to provide students with model sets to complete the review if it is assigned as a class activity.

The worksheet could be completed as a homework assignment.

Materials

- student worksheet
- transparency—Check-in
- wintergreen candy—one per student
- molecular model sets (optional)

LESSON 22 GUIDE

Who Nose?
Unit Review

Engage (5 minutes)

Key Question: How is smell related to molecular structure and properties?

> **ChemCatalyst**
> Name three items that might be on an exam covering the entire Smells Unit. Compose a question that could be included on the exam.

Sample Answers: Students might mention structural formulas, functional groups, the octet rule, Lewis dot symbols, electronegativity, polarity, electron domain theory, the receptor site model, and so on.

Discuss the ChemCatalyst

- Begin to review what was learned in the Smells Unit.
- Write general topics on the board as students share their answers. This becomes a review list.
- Ask students to save their exam questions for later in the class.

Sample Question

- What kinds of things might be on an exam for this unit?

Explore (20 minutes)
Introduce the Classwork

- Students can work in pairs to review concepts from Unit 2: Smells.

LESSON 22 CLASSWORK

Who Nose?
Unit Review

Name _____
Date _____ Period _____

Purpose
To integrate your learning about smell chemistry and to review the entire Smells unit.

Questions

1. Consider the molecular formula C_2H_4O. Draw two correct structural formulas for this molecular formula, then complete the rest of the table.

a. Structural formulas	H–C(H)(H)–C(=O)–H	H(H)C=C(H)–O–H
b. Polarity	polar	polar
c. Smell?	yes	yes
d. Reasons	Small polar molecules have a smell.	Small polar molecules have a smell.

2. Consider the three structural formulas in the table for the molecular formula $C_4H_{10}O$. Circle the alcohol, then complete the table.

Structural formulas	(H–C–C–C–C–O–H with H's) *circled*	H–C–C–O–C–C–H with H's	H–C–C–O–C–H with a C–H branch
a. Polarity	polar	polar	polar
b. Smell?	yes	yes	yes
c. Reasons	asymmetrical molecule, medium-size, an alcohol	asymmetrical molecule (bond around the O atom is bent), medium-size, an ether	asymmetrical molecule, medium-size, an ether

Imagine that the hydroxyl functional group in the alcohol is changed to an amine functional group. Draw the new structural formula. What would remain the same about the molecule, and what would change?

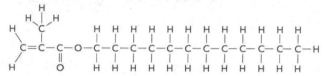

Same: Same number of carbon atoms. Similar shape—stringy. The molecule is still polar.

Different: There would be a nitrogen atom in the molecule, no oxygen atom, and an additional hydrogen atom. The structural formula and molecular formula would both change. The smell would change to fishy.

3. Lauryl methacrylate is a compound added to engine oil to increase its viscosity or thickness. Its molecular formula is $C_{16}H_{30}O_2$.

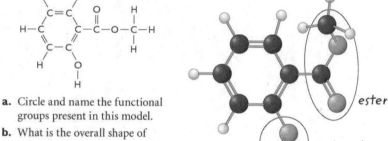

a. Predict the phase and polarity and whether the compound has a smell.

Phase: liquid   Polarity: polar   Smell: yes

b. Explain your reasoning for the answers to part a.

Phase: medium-size covalent molecule
Polarity: asymmetrical, has a polar end
Smell: medium-size, so it will smell

4. Consider the molecule $C_8H_8O_3$, called methyl salicylate. Its structural formula and three-dimensional structure are shown.

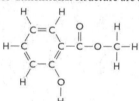

a. Circle and name the functional groups present in this model.

b. What is the overall shape of the molecule?

Frying pan-shaped; has a carbon ring

c. Predict the smell of methyl salicylate.   *minty and/or sweet*

d. Explain your reasoning.

Molecular shape points to either minty or sweet. Ester functional group points to sweet.

e. Does methyl salicylate have a mirror-image isomer? Explain.

No. There are no carbon atoms with four different atoms or groups of atoms attached.

Explain and Elaborate (15 minutes)
Go Over Answers to Review Exercises

- Ask students to come to the board to draw their structural formulas as time allows.
- Test students' predictions for methyl salicylate.
- Record the smell predictions of the class for methyl salicylate. You might want to have a ball-and-stick model of methyl salicylate available for inspection.
- Pass out one candy to each student for students to smell. *Note:* Remind students never to taste any substances in the lab.

Sample Questions

- What do you predict methyl salicylate will smell like? What is your reasoning?
- How does it actually smell?
- What smell categories does it fit into?
- Does the overall shape of the molecule fit with the smell? (both stringy and planar)

Key Point

The methyl salicylate molecule has structural features of both a minty molecule and a sweet molecule. Methyl salicylate is the compound in wintergreen candy that smells minty fresh and sweet. It is kind of frying-pan shaped and also has a long, stringy tail. It also has an ester functional group, like many sweet-smelling molecules.

Review the Key Chemistry Ideas Covered in the Smells Unit

- Continue to create a list of topics and terms on the board as a study guide.
- Ask students to share the exam questions they wrote for the ChemCatalyst question. Use these as review questions. Some sample questions are provided.

Sample Questions

- What information does a molecular formula contain?
- What does a molecular formula tell you about smell?
- Why do smell chemists use smell classifications?
- What is a structural formula? How do you draw a structural formula?

- What is a functional group? Name some. What can functional groups tell you about a molecule?
- What is the HONC 1234 rule, and where does it come from?
- What is a Lewis dot structure? How do you draw one?
- What are lone pairs and bonded pairs?
- Why are some molecules with three atoms bent while others are linear?
- How is three-dimensional molecular structure related to smell?
- What is a covalent bond? An ionic bond? A polar covalent bond?
- What is a dipole? How do you determine the direction of a dipole?
- What does polarity have to do with smell?
- What does the size of a molecule have to do with smell?
- What types of substances do not have a smell?
- Why does the Smells Unit deal mostly with molecules?

Evaluate (5 minutes)

Check-in

Would compounds made of either of these molecules have a smell? If so, try to predict what the smell would be. Explain your reasoning.

$$\begin{array}{c} \text{H H H H H H H H H O} \\ \text{H—C—C—C—C—C—C—C—C—C—C—Ö—H} \\ \text{H H H H H H H H H} \end{array} \quad \begin{array}{c} \text{H H} \\ \text{H—C—Ö—C—H} \\ \text{H H} \end{array}$$

Answer: Both molecules would have a smell. The first molecule (decanoic acid) is a medium-size carboxylic acid. Both its name and its structural formula indicate that it should smell putrid, which is its actual smell. The second molecule (dimethyl ether) is a small polar molecule, so we predict that it should have a smell. (Students have not smelled ether, so they probably will not be able to predict a smell for this compound.)

Homework

Assign the Section IV Review and the Unit Review for Unit 2: Smells in the student text.